How to . . .

get the most from your

C

Key Point

Basic concepts in point form.

Close Up

Additional hints, notes, tips or background information.

Watch Out!

Areas where problems frequently occur.

Quick Tip

Concise ideas to help you learn what you need to know.

Remember This!

Essential material for mastery of the topic.

COLES NOTES

How to get an *A* in . . .

Trigonometry & Circle Geometry

Cosines & sines

Trig functions

Solving equations

ABOUT COLES NOTES

COLES NOTES have been an indispensable aid to students on five continents since 1948.

COLES NOTES now offer titles on a wide range of general interest topics as well as traditional academic subject areas and individual literary works. All COLES NOTES are written by experts in their fields and reviewed for accuracy by independent authorities and the Coles Editorial Board.

COLES NOTES provide clear, concise explanations of their subject areas. Proper use of COLES NOTES will result in a broader understanding of the topic being studied. For academic subjects, Coles Notes are an invaluable aid for study, review and exam preparation. For literary works, COLES NOTES provide interesting interpretations and evaluations which supplement the text but are not intended as a substitute for reading the text itself. Use of the NOTES will serve not only to clarify the material being studied, but should enhance the reader's enjoyment of the topic.

Cataloguing in Publication Data

Bates, Bryce, 1952–

How to get an A in ... trigonometry & circle geometry :
cosines & sines, trig functions, solving equations

ISBN 0-7740-0556-4

1. Trigonometry. 2. Trigonometry–Problems, exercises, etc.
I. Title. II. Series.

QA531.B37 1998   516.24   C98-930275-X

Copyright 2008 and Published by
Coles Publishing
A division of Prospero Books
Toronto Canada
Publisher: Indigo Books and Music Inc.

Designed and Printed in Canada

Printed on Legacy Book Opaque 100%, manufactured from
100% post-consumer waste and is FSC-certified.
Manufacturing this book in Canada ensures compliance with strict
environmental practices and eliminates the need for international freight
shipping, a major contributor to global and air pollution.
Manufactured by Webcom

Contents

Trigonometry and circle geometry
Introduction

Imagine crossing a bridge that had not been properly engineered or living in a house with a roof which might collapse at any moment. Perhaps there is an old tree in your backyard or neighbourhood, with an "immeasurable" height that sparks your curiosity. Or perhaps you want to gauge how hard and what angle to hit a billiard ball. Each of these situations is connected to trigonometry.

Trigonometry is basically the branch of mathematics that deals with triangles. One of the earliest uses for trigonometry was back in 250 BC when Aristarchus used trigonometry to measure the relative distances of the sun and the moon from the Earth. Since these ancient times, the applications of trigonometry have extended to many fields that may not seem obvious to students.

A surveyor, for instance, uses triangulation to determine the shape, size and boundaries of a piece of land. These data can be used for settling legal disputes regarding property ownership as well as determine the market value of the land. Information gathered by a surveyor is also necessary to make decisions regarding construction at a given site.

The principles of trigonometry can be used quite reliably to measure heights or distances which are difficult to measure directly. The heights of trees, cliffs, buildings or aircraft and distances across bodies of water or between planets and stars can be calculated— all with the aid of trigonometry.

The concepts of trigonometry are applied extensively in the field of engineering when buildings, bridges and other free-standing structures are designed. As can be seen at virtually any building site, triangles are used widely in framing to maintain the sturdiness and provide the supporting strength necessary to the finished structure. Not only are they used in the construction of satellite dishes for stability with a lightweight frame but the successful transmission of communication signals also relies on trigonometric calculations.

Because of the repeating or periodic nature of the trigonometric functions, they are central to the study of water waves, musical sounds, alternating current and electromagnetic waves, including light, microwaves and radio waves. They are also used to describe all sorts of objects with circular motion, such as a Ferris wheel,s parts of a bicycle or pistons in an automobile. In addition, the ideas of trigonometry are often used in the interpretation of data obtained in the field of economics.

This is why we can say that trigonometry has proven itselfto be a fundamental branch of mathematics, enabling us both to understand the universe of which we are part and to make effective use of the resources in the world around us.

Other Coles Notes covering topics in senior mathematics:

Topics in senior math are frequently inter-connected at each grade level. These additional titles from Coles Notes will help you master them all:

How to Get an A in ...
• Permutations and Combinations
• Senior Algebra
• Sequences and Series
• Statistics and Data Analysis
• Calculus

CHAPTER ONE

Similar triangles

As you already know, all triangles have three sides. However, their shapes vary according to the angles formed at their vertices and the relative lengths of their sides. Similar triangles have the same shape but their sides are different lengths.

In any pair of similar triangles:
- all corresponding angles are equal.
- the ratios of the corresponding sides are equal.

The symbol ~ is used to indicate two triangles are similar to each other.

EXAMPLE 1

Given $\triangle ABC$ and $\triangle DEF$ below:

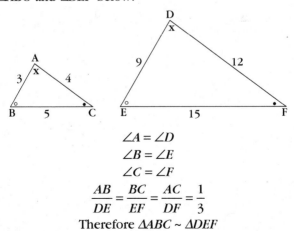

$$\angle A = \angle D$$
$$\angle B = \angle E$$
$$\angle C = \angle F$$
$$\frac{AB}{DE} = \frac{BC}{EF} = \frac{AC}{DF} = \frac{1}{3}$$

Therefore $\triangle ABC \sim \triangle DEF$

In order to determine whether or not two triangles are similar, we do not need to check all six pairs of measures (3 pairs of corresponding angles and 3 ratios of corresponding sides).

1

In the examples which follow, it is assumed that you are already familiar with angle theorems such as OAT (Opposite Angle Theorem), SATT (Sum of the Angles in a Triangle Theorem) and PLT (Parallel Line Theorem). However, in case you've forgotten, the ideas are summarized below.

OAT (OPPOSITE ANGLE THEOREM)
   If two lines intersect, then the opposite angles are equal.

SATT (SUM OF THE ANGLES IN A TRIANGLE THEOREM)
   The sum of the interior angles of a triangle is 180°. If two angles in one triangle are equal to two respective angles in another triangle, then the third angles are equal.

PLT (PARALLEL LINE THEOREM)
   When a transversal intersects two parallel lines, then
   • the alternate angles are equal.
   • the corresponding angles are equal.
   • the interior angles on the same side of the transversal are supplementary.

There are four theorems which can be used to prove similarity.

Angle-Angle-Angle similarity theorem (AAA~)

If the three angles of one triangle are equal to the three angles of another triangle, then the two triangles are similar.

EXAMPLE 2

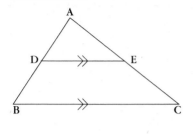

Given: $DE \parallel BC$
Prove $\triangle ABC \sim \triangle ADE$

Proof:

$\angle BAC = \angle DAE$ (common)
$DE \parallel BC$ (given)
$\angle ABC = \angle ADE$ (PLT)
$\angle ACB = \angle AED$ (PLT)
$\therefore \triangle ABC \sim \triangle ADE$ (AAA~)

Since the sum of the three angles in a triangle is always 180°(SATT), if a pair of triangles has two angles which are equal, then the third angles must be equal. Our first theorem, AAA ~ can be reduced to AA ~.

Angle-Angle similarity theorem (AA ~)

If two angles in one triangle are equal to two angles in another triangle, then the two triangles are similar.

EXAMPLE 3

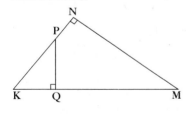

Given: $\angle KQP = 90°$
$\angle KNM = 90°$
Prove $\triangle KPQ \sim \triangle KMN$

Proof:

$\angle KQP = \angle KMN = 90°$(given)
$\angle QKP = \angle NKM$ (common)
$\therefore \triangle KPQ \sim \triangle KMN$ (AA~)

Remember that, in a formal proof, it is standard to write the names of the angles and sides in an order so that corresponding vertices match. For instance, it is considered better to write $\angle QKP = \angle NKM$ than to write $\angle QKP = \angle MKN$ because Q corresponds with N and P corresponds with M.

The other two similarity theorems involve corresponding sides.

Side-Angle-Side similarity theorem (SAS~)

If two triangles have ratios of two corresponding sides equal and the included angles equal, then they are similar.

EXAMPLE 4

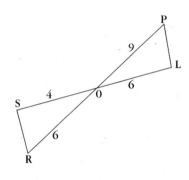

Given: PR intersects SL at O,
$LO = 6, PO = 9, RO = 6,$
$SO = 4$

Prove $\triangle STO \sim \triangle LPO$

Proof:

$$\frac{SO}{LO} = \frac{4}{6} = \frac{2}{3}$$

$$\frac{RO}{PO} = \frac{6}{9} = \frac{2}{3}$$

$$\therefore \frac{SO}{LO} = \frac{RO}{PO} \qquad \text{(equality)}$$

$$\angle SOR = \angle LOP \qquad \text{(OAT)}$$

$$\therefore \triangle SRO \sim \triangle LPO \qquad \text{(SAS~)}$$

Note again the matching of the corresponding vertices in the proof shown above.

Side-Side-Side similarity theorem (SSS~)

If all the ratios of corresponding sides are equal, then two triangles are similar.

EXAMPLE 5

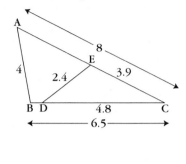

Given: $AB = 4, AC = 8, BC = 6.5,$
$\qquad CD = 4.8, CE = 3.9, DE = 2.4$

Prove $\triangle ABC \sim \triangle DEC$

Proof:

$$\frac{DE}{AB} = \frac{2.4}{4} = 0.6$$

$$\frac{EC}{BC} = \frac{3.9}{6.5} = 0.6$$

$$\frac{DC}{AC} = \frac{4.8}{8} = 0.6$$

$$\frac{DE}{AB} = \frac{EC}{BC} = \frac{DC}{AC} \quad \text{(equality)}$$

$$\therefore \triangle ABC \sim \triangle DEC \quad (SSS \sim)$$

Note that we could also use SAS $\sim$ in the previous example to prove the triangles are similar, since $\angle C$ is common to both triangles.

Now that we have discussed each of the similarity theorems, let's consider a situation where these ideas are extended.

EXAMPLE 6

a) Given: PM and GL
    are perpendicular to
    GP, $GP = 5$, $GL = 10$,
    $\angle PLG = \angle MGP$
    Prove $\triangle GPM \sim \triangle LPG$

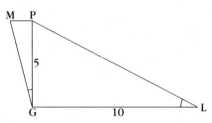

First we should study the diagram to decide which theorem would be suitable. Notice that we are only given one side (GP) of $\triangle GPM$. This rules out the SAS and SSS theorems. However, with the information indicated, we can use AA $\sim$.

Proof: $PM \perp GP$ (given)

$\qquad \angle GPM = 90°$ (definition of perpendicular)

$\qquad GL \perp GP$ (given)

$\qquad \angle LGP = 90°$ (definition of perpendicular)

$\qquad \angle GPM = \angle LPG$ (equality)

$\qquad \angle MGP = \angle PLG$ (given)

$\qquad \therefore \triangle GPM \sim \triangle LPG$ (AA~)

b) Find the length of *MP*.

Let $x = MP$ Then by similarity $\dfrac{LG}{GP} = \dfrac{PG}{x}$

Substituting $\dfrac{10}{5} = \dfrac{5}{x}$

Cross-multiplying $10x = 25$

Dividing $x = 2.5$

PRACTICE EXERCISES

1. Identify a pair of triangles in each of the diagrams and prove they are similar:

a)

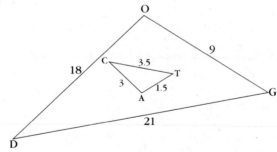

b)

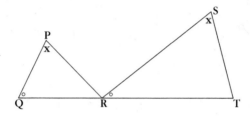

c)

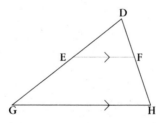

2   a)   Given: $UV \parallel XY$, UY intersects VX
         at Z, $UZ = 3$, $VZ = 2.25$,
         $YZ = 8$
         Prove $\triangle XYZ \sim \triangle VUZ$

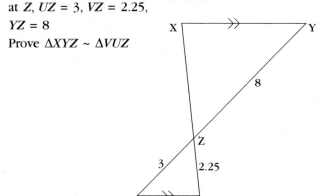

    b)   Find the length of XZ.

3   a)   Given: $\angle ADE = 50°$, $\angle AGB = 50°$,
         $AD = 15$, $AE = 10$, $EG = 14$
         Prove $\triangle BAG \sim \triangle EAD$

    b)   Find the length of AB.

CHAPTER TWO

Trigonometric ratios
of acute angles

As we have seen, the ratios of corresponding sides in similar triangles are equal. Other important properties of similar triangles lead us to trigonometry. It is convenient to relate the trigonometric ratios to the acute angles in a right triangle. So let's consider the following diagram containing two right triangles:

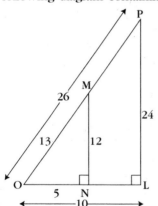

$\angle MON = \angle POL$ (common)
$\angle MNO = \angle PLO = 90°$ (given)

Therefore $\triangle MON \sim \triangle POL$ (AA~)
By similarity,
$$\frac{MN}{PL} = \frac{ON}{OL} = \frac{MO}{PO}$$

Other important equalities, relevant to trigonometry, can also be seen here. For example,

$$\frac{MN}{MO} = \frac{PL}{PO}\left(=\frac{12}{13}\right)$$

$$\frac{ON}{MO} = \frac{OL}{PO}\left(=\frac{5}{13}\right)$$

$$\frac{MN}{ON} = \frac{PL}{OL}\left(=\frac{12}{5}\right)$$

Each of these ratios has a special name depending on the angle associated with it. Since triangles can be oriented in different ways, the sides of the triangle are labelled, as shown below, in relation to the angle being considered.

8

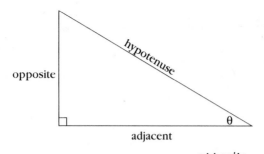

By definition: sine of θ (sinθ) $= \dfrac{opposite}{hypotenuse}$

cosine of θ (cosθ) $= \dfrac{adjacent}{hypotenuse}$

tangent of θ (tanθ) $= \dfrac{opposite}{adjacent}$

These are the 3 primary trigonometric ratios.

There are 3 reciprocal trigonometric ratios.

cosecant of θ (cscθ) $= \dfrac{1}{\sin\theta} = \dfrac{hypotenuse}{opposite}$

secant of θ (secθ) $= \dfrac{1}{\cos\theta} = \dfrac{hypotenuse}{adjacent}$

cotangent of θ (cotθ) $= \dfrac{1}{\tan\theta} = \dfrac{adjacent}{opposite}$

$\sin\theta = \dfrac{Opposite}{Hypotenuse}$	$\csc\theta = \dfrac{Hypotenuse}{Opposite}$
$\cos\theta = \dfrac{Adjacent}{Hypotenuse}$	$\sec\theta = \dfrac{Hypotenuse}{Adjancet}$
$\tan\theta = \dfrac{Opposite}{Adjacent}$	$\cot\theta = \dfrac{Adjacent}{Opposite}$

The acronym SOHCAHTOA (SINE - Opposite - Hypotenuse - COSINE - Adjacent - Hypotenuse - TANGENT - Opposite - Adjacent) is useful as a memory aid. Try pronouncing it as if SOHCAHTOA were a volcano in the South Seas.

EXAMPLE 1

Find the six trigonometric ratios of the angle indicated in the following triangle:

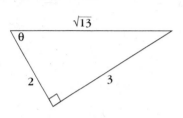

$$\sin\theta = \frac{3}{\sqrt{13}} \qquad \csc\theta = \frac{\sqrt{13}}{3}$$

$$\cos\theta = \frac{2}{\sqrt{13}} \qquad \sec\theta = \frac{\sqrt{13}}{2}$$

$$\tan\theta = \frac{3}{2} \qquad \cot\theta = \frac{2}{3}$$

Sometimes we need to use the Pythagorean relation ($a^2 = b^2 + c^2$) to calculate the third side of a right triangle before we can write the trig ratios.

EXAMPLE 2

Find the six trigonometric ratios of the angle indicated in the following triangle:

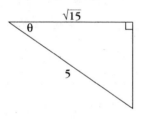

Let the unknown side's length be x.
Using the Pythagorean Theorem,

$$x = \sqrt{\left(5\right)^2 - \left(\sqrt{15}\right)^2}$$
$$x = \sqrt{25 - 15}$$
$$x = \sqrt{10}$$

Since x is opposite θ,

$$\sin\theta = \frac{\sqrt{10}}{5} \qquad \csc\theta = \frac{5}{\sqrt{10}}$$

$$\cos\theta = \frac{\sqrt{15}}{5} \qquad \sec\theta = \frac{5}{\sqrt{15}}$$

$$\tan\theta = \frac{\sqrt{10}}{\sqrt{15}} = \frac{\sqrt{6}}{3} \qquad \cot\theta = \frac{\sqrt{15}}{\sqrt{10}} = \frac{\sqrt{6}}{2}$$

We might encounter problems where one of the trigonometric ratios is given instead of the diagram of a triangle and we are asked to find the other five ratios. In these cases it is advisable to:

* draw a labelled diagram
* calculate the third side
* determine the required ratios.

EXAMPLE 3

If $\sin\alpha = \frac{3}{5}$ and α is acute, determine the other five trigonometric ratios.

Since we know $\sin\alpha$ then the side opposite this angle can be represented by 3 and the hypotenuse, by 5.

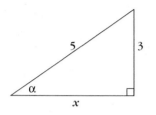

Let x = the adjacent side.
Using the Pythagorean Theorem,

$$x = \sqrt{5^2 - 3^2}$$
$$x = 4$$

Therefore $\sin\alpha = \dfrac{3}{5}$ $\csc\alpha = \dfrac{5}{3}$

$\cos\alpha = \dfrac{4}{5}$ $\sec\alpha = \dfrac{5}{4}$

$\tan\alpha = \dfrac{3}{4}$ $\cot\alpha = \dfrac{4}{3}$

There are two special triangles containing angles for which the primary trig ratios should be memorized.

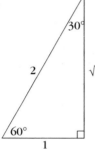

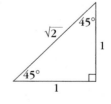

$\sin 60° = \dfrac{\sqrt{3}}{2}$ $\sin 30° = \dfrac{1}{2}$ $\sin 45° = \dfrac{1}{\sqrt{2}}$

$\cos 60° = \dfrac{1}{2}$ $\cos 30° = \dfrac{\sqrt{3}}{2}$ $\cos 45° = \dfrac{1}{\sqrt{2}}$

$\tan 60° = \sqrt{3}$ $\tan 30° = \dfrac{1}{\sqrt{3}}$ $\tan 45° = 1$

If you can at least remember the triangles, then the ratios can be determined easily. However, these two triangles are important ones to have "at your fingertips" because they are used to develop ratios for other angles, as we will see in Chapter 4. Also, they are often needed to solve other problems related to trigonometry.

EXAMPLE 4

Evaluate without using a calculator:

a) $\cos45°\cos30°+\sin45°\sin30°$

$$\cos45°\cos30° + \sin45°\sin30° = \left(\frac{1}{\sqrt{2}}\right)\left(\frac{\sqrt{3}}{2}\right) + \left(\frac{1}{\sqrt{2}}\right)\left(\frac{1}{2}\right)$$

$$= \frac{\sqrt{3}+1}{2\sqrt{2}}$$

b) $\sin^2 60°-\cot^2 60°$

$$\sin^2 60° - \cot^2 60° = \left(\frac{\sqrt{3}}{2}\right)^2 - \left(\frac{1}{\sqrt{3}}\right)^2$$

$$= \frac{3}{4} - \frac{1}{3}$$

$$= \frac{9}{12} - \frac{4}{12}$$

$$= \frac{5}{12}$$

Note that in (b), the exponent is written above the trig name, not the angle. This tells us to find the trig ratio of the angle given and then square the answer. For trigonometric functions this is the standard way of writing powers. You may, if you wish, rewrite. For example, $\sin^2 60°$ can be written as $(\sin 60°)^2$. However, suppose we were given $\sin(60°)^2$ instead. Then we would be expected to square the angle (i.e. $3600°$) and then find the sine ratio.

Of course, there are other acute angles between $0°$ and $90°$ besides the three we have just considered. The easiest way of evaluating these ratios is with the use of a scientific calculator. The order of buttons that you need to push depends on the brand of calculator you are using. See the manufacturer's manual, if necessary.

The calculator allows us to directly evaluate the primary trig ratios for any angle. If we need to determine any of the reciprocal trig ratios, we can find the value of the corresponding primary ratio and then use the 1/x button.

Be sure your calculator is in the DEG (Degree) mode first.

EXAMPLE 5

Find the following trigonometric values, rounded to 4 decimal places:

a) $\sin 25°$

$\sin 25° = 0.4226$

b) $\tan 18°$

$\tan 18° = 0.3249$

c) $\csc 40°$

$\sin 40° = 0.6427876 \left(= x\right)$

$\csc 40° = \dfrac{1}{0.6427876} \left(= \dfrac{1}{x}\right)$

$\csc 40° = 1.5557$

d) $\cot 4.9°$

$\tan 4.9° = 0.0857302 \left(= x\right)$

$\cot 4.9° = \dfrac{1}{0.0857302} \left(= \dfrac{1}{x}\right)$

$\cot 4.9° = 11.6645$

The calculator can also be used to do the reverse operation. That is, we may be given the value of the trig ratio and use it to find the angle. Here, we use the(inverse) functions which are generally accessed by pressing the SHIFT/2ND FUNC key followed by the corresponding trig keys. If we need the angle for csc, sec or cot, we find the corresponding primary trig ratio first using the 1/x button and then use the SHIFT and appropriate trig key to get the angle.

EXAMPLE 6

Find the measure of the angle to the nearest tenth of a degree:

a) $\sin\theta = 0.9063$

$\theta = 65.0°$

b) $\cos\beta = 0.7869$

$\beta = 38.1°$

c)   csc⌀ = 1.1547   d)   cotε = 1.2349

$$\sin\phi = \frac{1}{1.1547} = 0.8660258 \qquad \tan\varepsilon = \frac{1}{1.2349} = 0.8097821$$

$\phi = 60.0°$ $\varepsilon = 39.0°$

PRACTICE EXERCISES

1.   Find the primary trigonometric ratios for the angle indicated:

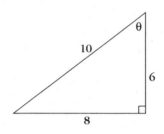

2.   Find the six trigonometric ratios for the angle indicated:

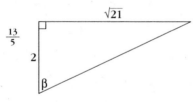

3.   If secθ =   and θ is acute, determine the values of the other five trigonometric ratios.

4.   Without using a calculator, evaluate:
    a) tan45°   b) cos60°   c) csc45°
    d) $\sin30°\sin^2 60°$ e) sin45°cos45° + cos30°tan60°

5.   Find the following trigonometric values, rounded to 4 decimal places:
    a) sin3°   b) cos82.9°   c) tan34°
    d) cot76°   e) sec83.5°   f) csc12.3°

6.   Find the measure of the angle to the nearest tenth of a degree:
    a) cosθ = 0.6032 b) sinα = 0.0951   c) tanλ = 2.2460
    d) secμ = 1.0403 e) cscη = 2   f) cotδ = 0.0248

Solving right-angled triangles

To solve a triangle means to determine the measures of any angles and sides not already given. In this chapter, we will consider only right triangles. Essentially, there are three tools/ideas we need to use here:

- SATT
- Pythagorean Theorem
- Trigonometric Ratios

There are two basic types of situations we can encounter:
1)   Given one side and one angle other than the right angle.
2)   Given two sides and the right angle.

Before we study examples of these, there is a convention of naming sides with which you should be familiar.

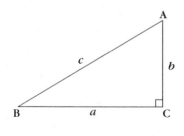

Although the two-letter mode of naming sides is widely used, a small case letter for the side opposite a given vertex is also acceptable. See the diagram to the left. Instead of referring to side AB, we can call it c since it is opposite vertex C.

EXAMPLE 1
Solve $\triangle TUV$, correct to one decimal place, given $\angle U = 90°$, $\angle T = 34°$, $v = 12$.

Let's start by sketching the triangle, labelling the known values.

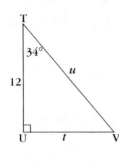

Using SATT, we can write

$$\angle V = 90° - 34°$$
$$= 56°$$

Now, we can find either t or u by using an appropriate trig ratio. Let's solve for t.

Since v is adjacent to $\angle T$ and t is opposite $\angle T$, then

$$\frac{t}{v} = \tan 34°$$

$$\therefore t = 12\tan 34°$$
$$= 12(0.6745085)$$
$$= 8.1$$

To find u, we can either use another trig ratio or the Pythagorean Theorem.

Since u is the hypotenuse,

$$\frac{v}{u} = \cos 34°$$

$$\therefore u = \frac{v}{\cos 34°}$$

$$= \frac{12}{0.8290375}$$

$$= 14.5$$

$$[\text{or } u^2 = t^2 + v^2$$
$$= (8.1)^2 + (12)^2$$
$$= 209.61$$

$$\therefore u = \sqrt{209.61}$$
$$= 14.5$$

Thus, we have completed our solution of ΔTUV. Trig ratios containing $\angle V$ instead of $\angle T$ can be used to solve for t and/or u if you prefer.

EXAMPLE 2

Solve ΔPQR, rounded to one decimal place, given $\angle Q = 90°$, $p = 5$, $q = 8$.

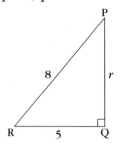

Using the Pythagorean Theorem,

$$r^2 = q^2 - p^2$$
$$= 8^2 - 5^2$$
$$= 39$$
$$\therefore r = \sqrt{39}$$
$$= 6.2$$

We can use a trig ratio to find $\angle P$ or $\angle R$. Suppose we choose to find $\angle R$ first. Using the sides given,

$$\cos R = \frac{5}{8}$$
$$= 0.625$$

$$\therefore \angle R = 51.3° \quad \text{(using } \cos^{-1})$$
$$\text{Using SATT, } \angle P = 90° - 51.3°$$
$$= 38.7°$$

In the above examples, note that all of the calculator digits are kept until the final rounding of the answers. Also, in order to minimize the introduction of errors caused by rounding, we try, where possible, to use the given information for our calculations.

Often, there are situations where we only need to determine one particular side or angle, rather than solving the whole triangle.

EXAMPLE 3

Determine, correct to one decimal place, the length of x:

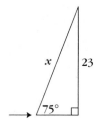

Since 23 is the measure of the side opposite the 75°angle, and x is the hypotenuse,

$$\frac{23}{x} = \sin 75°$$

$$\therefore x = \frac{23}{\sin 75°}$$

$$= \frac{23}{0.9659258}$$

$$= 23.8$$

EXAMPLE 4

Determine, correct to one decimal place, the measure of θ.

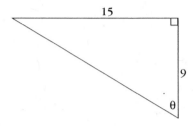

The side opposite θ is 15 while the side adjacent is 9.

$$\therefore \tan\theta = \frac{15}{9}$$
$$= 1.6$$
$$\theta = 59.0°$$

There are many practical problems involving right triangles which can be solved using trig ratios. The terms angle of elevation and angle of depression, illustrated below, are often used to convey known information in a simple way.

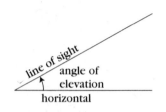

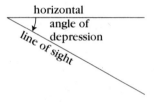

EXAMPLE 5

From a lighthouse window, David sights a sailboat at an angle of depression of 43°. If his eyes are 39 m above the water, how far is the sailboat from the base of the lighthouse?

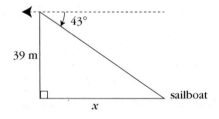

We are asked to find the horizontal distance from the boat to the lighthouse. Assuming the lighthouse is standing upright, we can draw a right triangle labelling the vertical leg as 39 m and the horizontal leg with a variable, like *x*.

Since the angle of depression is always measured down from the horizontal, we can mark it on the diagram as shown. Because this horizontal is parallel to x, the angle between x and the hypotenuse is also 43°.

$$\therefore \frac{39}{x} = \tan 43°$$
$$\therefore x = \frac{39}{\tan 43°}$$
$$= \frac{39}{0.932515}$$
$$= 41.8$$

Therefore, the sailboat is 41.8 m away from the base of the lighthouse.

Don't forget that word problems should be completed with a final statement which answers the question asked including appropriate units of measurement.

Problems using trig ratios often involve calculating distances which are difficult to measure directly. Sometimes two or more triangles may be needed to solve a given problem.

EXAMPLE 6

A statue 8 m tall sits on top of a hill. To an observer standing on level ground at the bottom of the hill, the angle of elevation of the top of the statue is 25° and the angle of elevation of its base is 17°. How far above ground level is the top of the statue?

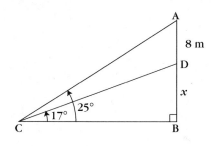

Based on the information we are given, we draw a diagram and label it as shown. To answer the question, we need to find the height of AB. Since we already know AD (the height of the statue), we must determine DB or x.

Let x be the vertical height of the hill.

We do not have enough information in $\triangle DBC$ to calculate x directly. However, both of the right triangles, $\triangle ABC$ and $\triangle DBC$, share side CB.

$$\tan 17° = \frac{x}{CB} \text{ and } \tan 25° = \frac{8+x}{CB}$$

19

Since we are interested in finding x, not CB, we rearrange these equations and equate them to eliminate CB.

$$CB = \frac{x}{\tan 17^\circ} \qquad CB = \frac{8+x}{\tan 25^\circ}$$

$$\frac{x}{\tan 17^\circ} = \frac{8+x}{\tan 25^\circ}$$

Now, cross-multiplying and solving for x,

$$x\tan 25^\circ = \left(8 + x\right)\tan 17^\circ$$

$$x\left(\tan 25^\circ - \tan 17^\circ\right) = 8\tan 17^\circ$$

$$x = \frac{8\tan 17^\circ}{\tan 25^\circ - \tan 17^\circ}$$

$$= \frac{2.4458455}{0.1605769}$$

$$= 15.2$$

$$8 + 15.2 = 23.2$$

Therefore the top of the statue is 23.2 m above ground level.

EXAMPLE 7

To find the distance across a canyon, an engineer takes a sighting from a point A to a point C across the canyon and then, a sighting from B to C. Points A and B are 65 m apart. The angle measurements are shown on the diagram below. Find the perpendicular distance, d, across the canyon.

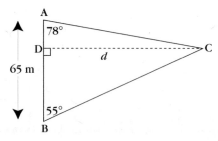

As we can see in the diagram, there is not enough information in either of the small triangles to find d directly. We don't know the measures of any of the sides in these triangles. However, we do know $AB = 65$ m. We can also see that $AD + DB = AB$.

From $\triangle ACD$, $\tan 78° = \dfrac{d}{AD}$

$\therefore AD = \dfrac{d}{\tan 78°}$

From $\triangle BCD$, $\tan 55° = \dfrac{d}{DB}$

$\therefore DB = \dfrac{d}{\tan 55°}$

By substitution, we get an equation containing the one unknown we want, d:

$$AD + DB = AB$$

$$\frac{d}{\tan 78°} + \frac{d}{\tan 55°} = 65$$

Now, we rearrange and solve for d.

$$\frac{d\tan 55° + d\tan 78°}{\tan 78° \tan 55°} = 65$$

$$d(\tan 55° + \tan 78°) = 65 \tan 78° \tan 55°$$

$$d = \frac{65 \tan 78° \tan 55°}{\tan 55° + \tan 78°}$$

$$= \frac{436.72903}{6.1327781}$$

$$= 71.2$$

Therefore, the distance across the canyon is 71.2 m.

Remember that we have defined the trig ratios in relation to right triangles. So, in the solutions for the problems illustrated above, all equations have been set up for right triangles only. We will deal with oblique or non-right triangles in the next chapter.

PRACTICE EXERCISES

1. Solve the following triangles, correct to one decimal place:
   a) given $\angle B = 90°$, $\angle C = 80°$, $b = 45$.
   b) given $\angle M = 90°$, $k = 9.3$, $m = 23.9$.

2. Given $\triangle XYZ$ with $\angle Z = 90°$, $\angle X = 32°$, $x = 20$ cm, find the length of y to the nearest centimetre.

3. A flagpole casts a shadow of 28 m when the angle of elevation of the sun is 35°. Find the height of the flagpole, rounded to one decimal place.

4. From the top of one building, the angles of depression of the top and bottom of a second building are 24.5° and 62.3° respectively. If the two buildings are 15 m apart, find their heights.

5. A man standing at a point west of a tree, finds the angle of elevation to its top is 26°. From a point east of the tree, he measures an angle of elevation of 29°. If the distance between the two points is 38 m, calculate the height of the tree.

CHAPTER FOUR

Trigonometric ratios of other angles

So far, we have considered trig ratios for acute angles as found in right triangles. We can extend these ideas to all angles, including negative angles using a set of coordinate axes.

The positive x-axis is the initial arm.

The line containing $P(x, y)$ is the terminal arm.

θ or $\angle POA$ is the angle of rotation. It is positive and increases in the counter-clockwise position. This angle is said to be in standard position.

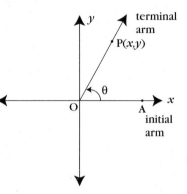

By starting with the initial arm and rotating it all the way around to the initial position, a circle is formed, with radius r having the general equation:

$$x^2 + y^2 = r^2$$

Notice that for almost any point, P, on the circle, a right triangle can be drawn by joining P to the x-axis.

The x-coordinate of P represents the side adjacent to θ while the y-coordinate represents the opposite side. The radius, r, is always positive and is the hypotenuse of such a triangle. As a result, trig ratios can be stated in terms of x, y, and r:

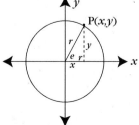

$$\sin\theta = \frac{y}{r} \qquad \csc\theta = \frac{r}{y}$$

$$\cos\theta = \frac{x}{r} \qquad \sec\theta = \frac{r}{x}$$

$$\tan\theta = \frac{y}{x} \qquad \cot\theta = \frac{x}{y}$$

EXAMPLE 1

a)

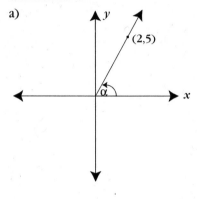

Find the primary trig ratios of the angle α whose terminal arm passes through (2, 5).

Since
$$x^2 + y^2 = r^2$$
$$2^2 + 5^2 = r^2$$
$$r^2 = 29$$
$$\therefore r = \sqrt{29}$$

$$\therefore \sin\alpha = \frac{5}{\sqrt{29}}$$

$$\cos\alpha = \frac{2}{\sqrt{29}}$$

$$\tan\alpha = \frac{5}{2}$$

b)

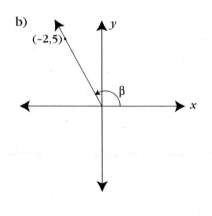

Find the primary trig ratios of the angle ß whose terminal arm passes through (-2, 5).

$$(-2)^2 + 5^2 = r^2$$
$$\therefore r = \sqrt{29}, (r < 0)$$

$$\therefore \sin\beta = -\frac{5}{\sqrt{29}}$$

$$\cos\beta = -\frac{2}{\sqrt{29}}$$

$$\tan\beta = -\frac{5}{2}$$

Note that in (b), if a vertical line were drawn from the point (-2, 5) to the x-axis, a right triangle would be formed. The angle between the terminal arm and the negative x-axis would be equal in size to α from (a) since (-2, 5) is a reflection of (2, 5) in the y-axis. This connection is important but, don't forget we are relating the trig ratios to angles formed by swinging the initial arm so that a circular path is created. The sides of a triangle would not have any negative values while the coordinates of points on the Cartesian grid can.

With the special triangles discussed in Chapter 2, we can develop relationships allowing us to find the trig ratios for many other angles. Let's begin by superimposing one of the special triangles on a set of coordinate axes:

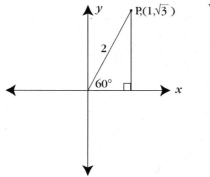

We already know

$$\sin 60° = \frac{\sqrt{3}}{2}$$

$$\cos 60° = \frac{1}{2}$$

$$\tan 60° = \sqrt{3}$$

Let's rotate the terminal arm into the second quadrant so $P_2(-1, \sqrt{3})$ is a reflection of $P_1(1, \sqrt{3})$

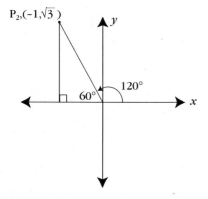

Since the x-axis is a straight line and hence, forms a 180° angle, the angle formed by the terminal arm is 180° - 60° or 120°.

$$\therefore \sin 120° = \frac{\sqrt{3}}{2}$$

$$\cos 120° = -\frac{1}{2}$$

$$\tan 120° = -\sqrt{3}$$

Repeating this rotation into the third quadrant gives us $P_3(-1,-\sqrt{3})$.

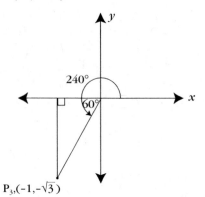

The new angle of rotation is $180° + 60°$ or $240°$.

$$\therefore \sin 240° = -\frac{\sqrt{3}}{2}$$
$$\cos 240° = -\frac{1}{2}$$
$$\tan 240° = \sqrt{3}$$

$P_3,(-1,-\sqrt{3})$

Finally, rotating into the fourth quadrant we reach $P_4(1,-\sqrt{3})$.

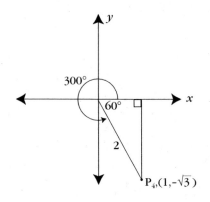

Here the angle of rotation is $360° - 60°$ or $300°$.

$$\therefore \sin 300° = -\frac{\sqrt{3}}{2}$$
$$\cos 300° = \frac{1}{2}$$
$$\tan 300° = -\sqrt{3}$$

$P_4,(1,-\sqrt{3})$

There are two important ideas illustrated above.

1. All the ratios for any first quadrant angle (like 60°) are positive.
   In the second quadrant, only Sine is positive.
   In the third quadrant, only Tangent is positive.
   In the fourth quadrant, only Cosine is positive.

   As indicated, this is referred to as the CAST rule.

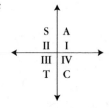

Some people prefer to use fun sayings like *All Soup Turns Cold* to help memorize this rule.

Note that each reciprocal trig ratio has the same sign (+/-) as its respective primary ratio in a given quadrant.

EXAMPLE 2

Angle θ is in the fourth quadrant and $\cos\theta = \frac{3}{5}$. Find the other trig ratios.

Sketch a labelled diagram:

From the diagram, we see y must be negative.

$$x^2 + y^2 = r^2$$
$$3^2 + y^2 = 5^2$$
$$y^2 = 25 - 9$$
$$= 16$$
$$\therefore y = -4$$

$$\sin\theta = -\frac{4}{5} \qquad \csc\theta = -\frac{5}{4}$$

$$\cos\theta = \frac{3}{5} \qquad \sec\theta = \frac{5}{3}$$

$$\tan\theta = -\frac{4}{3} \qquad \cot\theta = -\frac{3}{4}$$

2. As an extension of the CAST rule, we can develop some general relationships called the related acute angle formulas.

$$\text{As we saw,} \qquad \sin 120^\circ = \sin(180^\circ - 60^\circ)$$
$$= \sin 60^\circ$$
$$= \frac{\sqrt{3}}{2}$$
$$\cos 120^\circ = \cos(180^\circ - 60^\circ)$$
$$= -\cos 60^\circ$$
$$= -\frac{1}{2}$$

Let a be an acute angle (i.e. in the first quadrant).

For angles in the Second Quadrant

$$\sin (180° - a) = \sin a$$
$$\cos (180° - a) = - \cos a$$
$$\tan (180° - a) = - \tan a$$

For angles in the Third Quadrant

$$\sin (180° + a) = - \sin a$$
$$\cos (180° + a) = - \cos a$$
$$\tan (180° + a) = \tan a$$

For angles in the Fourth Quadrant

$$\sin (360° - a) = - \sin a$$
$$\cos (360° - a) = \cos a$$
$$\tan (360° - a) = - \tan a$$

The CAST rule will help you memorize these formulas. For instance, $\cos (180° - a) = -\cos a$ because cosine is negative in the second quadrant.

Use the following steps to calculate trig ratios of angles in the second, third or fourth quadrants:

* Determine in which quadrant the angle is located.
* Use the appropriate formula.
* Determine a by subtraction.
* Use a special triangle to evaluate the ratio.

In order to perform the first step, we need to know the measures of the angles along the axes. These are marked on the grid below:

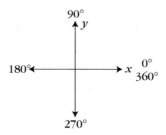

EXAMPLE 3

Evaluate without using a calculator:

a) $\cos 150°$ b) $\tan 225°$

a)   150° is in the second quadrant.

$$\therefore \cos 150° = \cos (180° - a)$$
$$= \cos (180° - 30°)$$
$$= -\cos 30°$$
$$= -\frac{\sqrt{3}}{2}$$

b)   225° is in the third quadrant.

$$\therefore \tan 225° = \tan(180° + a)$$
$$= \tan(180° + 45°)$$
$$= \tan 45°$$
$$= 1$$

Since angles on the coordinate axes are formed by rotating a terminal arm, there is no reason why we can't rotate beyond 360° Angles, like those shown below, which have the same terminal arm, are called coterminal angles.

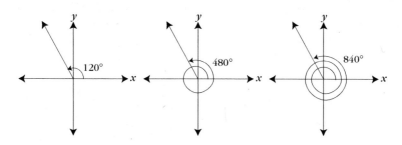

The trig ratios for coterminal angles are equal. The smallest positive angle between 0° and 360° is referred to as the related principal angle. In this case, it is 120°. The others are found by adding multiples of 360° to the related principal angle (i.e. 360° + 120° = 480°, 720° + 120° = 840°).

To determine trig ratios of a given coterminal angle, we divide the angle by 360°. The remainder is the related principal angle. From here, we proceed as in Example 3.

EXAMPLE 4

Evaluate:

    a) $\sin 405°$ b) $\cos 1320°$

    a) $405° \div 360° = 1 + 45°$ remainder

$$\therefore \sin 405° = \sin 45°$$
$$= \frac{1}{\sqrt{2}}$$

    b) $1320° \div 360° = 3 + 240°$ remainder

$$\therefore \cos 1320° = \cos 240°$$
$$= \cos (180° + 60°)$$
$$= -\cos 60°$$
$$= -\frac{1}{2}$$

If the terminal arm is rotated clockwise, instead of counter-clockwise, from the initial arm, then the angles produced are considered negative as shown below.

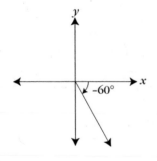

Each negative angle has a positive related principal angle. For the -60° angle shown at left, the related principal angle is 300°. Even though the relationship between a negative angle and its related principal angle can be used to evaluate trig ratios of negative angles, it is easier to use the following formulas.

Negative Angle Formulas

$$\sin(-a) = -\sin a$$
$$\cos(-a) = \cos a$$
$$\tan(-a) = -\tan a$$

An easy way to memorize these, is to note that the cosine for a negative angle is the same as the cosine for an equal positive angle.

Once again, the applications of these formulas are related to the use of the special triangles. This further expands the range of angles for which trig ratios can be calculated without the need for a calculator.

EXAMPLE 5

Evaluate without using a calculator:

    a) $\cos(-135°)$ b) $\tan(-300°)$ c) $\csc(-585°)$

    a) $\cos(-135°)$ $= \cos 135°$
$$= \cos(180° - 45°)$$
$$= -\cos 45°$$
$$= -\frac{1}{\sqrt{2}}$$

    b) $\tan(-300°)$ $= -\tan 300°$
$$= -\tan(360° - 60°)$$
$$= -(-\tan 60°)$$
$$= \sqrt{3}$$

    c) $-585° \div 360° = -1 + (-225°)$ remainder
$$\therefore \csc(-585°) = \frac{1}{\sin(-225°)}$$
$$= \frac{1}{-\sin 225°}$$
$$= \frac{1}{-\sin(180° + 45°)}$$
$$= \frac{1}{-(-\sin 45°)}$$
$$= \frac{1}{\frac{1}{\sqrt{2}}}$$
$$= \sqrt{2}$$

31

Notice in Example 5 that after we convert the negative angle to a positive one, we then consider the quadrant in which the positive angle is located and continue as usual.

In our discussion, we have focused on the use of the special triangles for evaluating trig ratios because, as mentioned before, the 30°, 45° and 60° ratios are used frequently. Of course, a calculator or computer can be used to determine, to a high degree of accuracy, the value for the trig ratio of any angle (or the angle of a given trig ratio) by using the same method described in Chapter 2.

EXAMPLE 6

a) Evaluate, using a calculator, answers rounded to four decimal places:

i) $\sin 234°$

$\sin 234° = -0.8090$

ii) $\cot(-871.2°)$

$$\cot(-871.2°) = \frac{1}{\tan(-871.2°)}$$

$$= 1.8190$$

b) Determine the angle, rounded to 1 decimal place:

i) $\cos\theta = -0.4317$

$\theta = 115.6°$

ii) $\csc^{-1}(-10.9)$

$$\csc^{-1}(-10.9) = \sin^{-1}\left(\frac{1}{-10.9}\right)$$

$$= -5.3°$$

Note that in Example 6(b), there are other possible answers not given by the calculator. For reasons we will deal with later, the calculator $\sin^{-1}$ and $\tan^{-1}$ functions only give angles between -90° and 90°, while the $\cos^{-1}$ function only provides angles between 0° and 90°. If we are interested in finding angles other than those provided by the calculator, other steps are necessary.

EXAMPLE 7

Given $\cos\theta = -0.8425$, find the value of θ, rounded to one decimal place, if it is located in the third quadrant and $0° \leq \theta \leq 360°$.

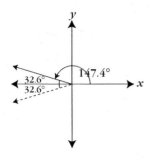

From the calculator, we get θ = 147.4° which is in the second quadrant. 180° - 147.4° = 32.6° as marked in the diagram. If we reflect the terminal arm into the third quadrant, as we did earlier when developing the CAST rule, then the angle rotated becomes 180° + 32.6° or 212.6°.

If you check cos 212.6° with your calculator, you will notice the answer is not exactly -0.8425. A small degree of error is acceptable here, since the first angle 147.4° has been rounded off.

EXAMPLE 8

Given cotθ = -0.1913, find two possible values of θ where 0° ≤ θ ≤ 360°.

$$\tan \theta = \frac{1}{-0.1913}$$
$$= -5.2274$$
∴ θ = -79.2° (reading from the calculator)

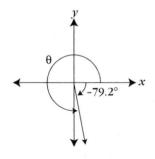

With the aid of a diagram, we get one required value by subtraction: 360° -79.2° or 280.8°

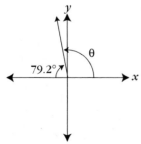

Using the CAST rule, we know there is another standard angle in the second quadrant for which the tangent ratio is negative. By reflecting -79.2° into the second quadrant, we get another angle: 180° -79.2° or 100.8°.

PRACTICE EXERCISES

1. Find the six trig ratios of the angle θ whose terminal arm passes through (-4, -7).

2. Angle ß is in the third quadrant and tanß = $\frac{5}{12}$. Find the other trig ratios for ß.

3. Evaluate without using a calculator:
   a) sin 135°   b) cos 135°   c) tan 150°
   d) cos 240°   e) cot 210°   f) sec 315°
   g) sin 390°   h) tan 780°   i) csc 420°
   j) cos 480°   k) cot 1305°   l) sin 1050°
   m) sin (-45°)   n) cos (-150°)   o) tan (-240°)
   p) sec (-300°)   q) cot (-135°)   r) csc (-930°)

4. Evaluate, rounded to four decimal places:
   a) cos 112°   b) tan 220°   c) sec (-134.5°)
   d) cot 573°   e) sin (-1130.2°)   f) csc 731.3°

5. Find the value of θ, rounded to one decimal place:
   a) sin θ = -0.1039   b) tan θ = -3.7028   c) sec θ = -4.0147

6. Find the value of θ, rounded to one decimal place, in the given quadrant if
   a) cos θ = 0.8414, fourth quadrant
   b) sin θ = -0.3907, third quadrant
   c) cot θ = -0.2536, second quadrant

7. Find two possible values of θ, rounded to one decimal place, where 0° ≤ θ ≤ 360°:
   a) sin θ = -0.6429   b) sec θ = -2.4188

CHAPTER FIVE

The law of cosines and the law of sines

In Example 7 of Chapter 3, we worked through an engineer's problem of determining the distance across a canyon. As we saw, the diagram consisted of a large oblique triangle with a perpendicular line drawn to represent the desired distance. This happened to yield two right triangles which enabled us to solve the problem. Suppose, instead, we had an oblique triangle but, were required to find the length of one its sides as opposed to its altitude?

EXAMPLE 1
Given $\triangle ABC$ with $\angle C = 37°$, $a = 11$, $b = 9.1$, calculate c correct to one decimal place.

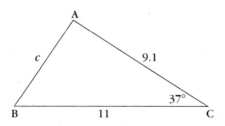

This is not necessarily a right triangle. So, it is incorrect to use trig ratios or the Pythagorean Theorem like we did in Chapter 4 to solve this problem.

However, to start, we might draw the altitude AD to create two right triangles.

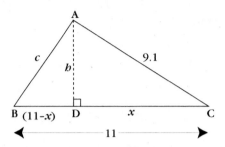

To simplify the discussion, we'll use single letters as labelled in the diagram.

Using the Pythagorean Theorem for $\triangle ABD$,
$$c^2 = (11 - x)^2 + h^2$$
$$= (11)^2 - 2(11)x + x^2 + h^2$$

But, for $\triangle ABD$, $\qquad (9.1)^2 = x^2 + h^2$

By substitution, $\qquad c^2 = (11)^2 - 2(11)x + (9.1)^2$
$$= (11)^2 + (9.1)^2 - 2(11)x$$

But in $\triangle ADC$, $\qquad \dfrac{x}{9.1} = \cos 37°$
$$\therefore x = 9.1\cos 37°$$

By substitution, $c^2 = (11)^2 + (9.1)^2 - 2(11)(9.1)\cos 37°$
$$= 121 + 82.81 - 159.88683$$
$$= 43.92317$$
$$\therefore c = 6.6$$

In the above example, $a = 11$, $b = 9.1$, $\angle C = 37°$. If we substitute the numbers with the variables, we obtain the general formula called the Cosine Law.

The Cosine Law

$$c^2 = a^2 + b^2 - 2ab\cos C$$

Two other forms of this law are commonly stated:
$$a^2 = b^2 + c^2 - 2bc\cos A$$
$$b^2 = a^2 + c^2 - 2ac\cos B$$

However, since the names of the vertices in a triangle can vary, it is more important to know one form of the law and when and how to use it.

As an aid to memorizing this formula, it is worth noting:
- its similarity to the Pythagorean Theorem
- the side to the left of the equal sign is opposite the angle for which you write the cosine.

EXAMPLE 2

Given $\angle X = 48°$, $y = 8.0$, $z = 5.0$, find x.

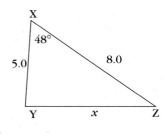

As usual, a labelled diagram should be drawn. But, try not to draw a right triangle here. You may find it helpful to start by roughly sketching the given angle at X and drawing the given sides so y is bigger than z. Then complete the triangle as shown to the left.

Applying the Cosine Law,

$$x^2 = y^2 + z^2 - 2yz\cos X$$
$$= (8.0)^2 + (5.0)^2 - 2(8.0)\,(5.0)\,\cos 48°$$
$$= 35.469551$$
$$\therefore x = 6.0$$

 Don't forget to use the proper order of operations here. Many students would do all the numerical operations but finish by multiplying everything by cos 48°.
This is wrong!

As a rough check, we can refer back to our diagram to see if the calculated value for x is reasonable.

The Cosine Law can also be used to find an unknown angle in a triangle if all the sides are known. Again, we assume, unless otherwise indicated, that it is not a right triangle.

EXAMPLE 3

Given $a = 10$, $b = 8$, $c = 17$, find $\angle C$.

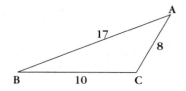

A rough sketch representing the relative measurements of the sides indicates $\angle C$ should be obtuse (> 90°).

Using the Cosine Law,

$$c^2 = a^2 + b^2 - 2ab\cos C$$

Rearranging $\dfrac{c^2 - a^2 - b^2}{-2ab} = \cos C$

Substituting $\cos C = \dfrac{(17)^2 - (10)^2 - (8)^2}{-2(10)(8)}$

$$= -0.78125$$

$$\therefore \angle C = 141.4°$$

Sometimes, depending on what is known, we must use another formula for solving oblique triangles.

Consider $\triangle ABC$ with altitude AD below:

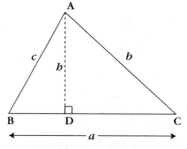

Using lower case letters for easy reference,

From $\triangle ABD$, $\sin B = \frac{h}{c}$

From $\triangle ACD$, $\sin C = \frac{h}{b}$

$\therefore h = b\sin C$

By substitution, $b\sin C = c\sin B$

Dividing both sides by $\sin B \sin C$ and simplifying, $\dfrac{b}{\sin B} = \dfrac{c}{\sin C}$

By drawing perpendiculars to the other sides of $\triangle ABC$ and repeating the above steps we obtain the complete Sine Law.

The Sine Law

$$\frac{a}{\sin A} = \frac{b}{\sin B} = \frac{c}{\sin C}$$

38

EXAMPLE 4

In $\triangle ABC$, $\angle A = 50°$, $\angle C = 75°$, and $a = 40.3$. Find sides b and c, rounded to one decimal place.

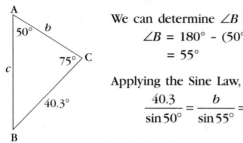

We can determine $\angle B$ here using SATT:
$$\angle B = 180° - (50° + 75°)$$
$$= 55°$$

Applying the Sine Law,
$$\frac{40.3}{\sin 50°} = \frac{b}{\sin 55°} = \frac{c}{\sin 75°}$$

Then, solving using pairs of ratios,

$$\frac{40.3}{\sin 50°} = \frac{b}{\sin 55°} \qquad\qquad \frac{40.3}{\sin 50°} = \frac{c}{\sin 75°}$$

$$\therefore b = \frac{40.3\sin 55°}{\sin 50°} \qquad\qquad \therefore c = \frac{40.3\sin 75°}{\sin 50°}$$

$$= \frac{40.3(0.819152)}{0.7660444} \qquad\qquad = \frac{40.3(0.9659258)}{0.7660444}$$

$$= 43.1 \qquad\qquad\qquad = 50.8$$

Note that we couldn't have used the Cosine Law here, since we only were given one side.

EXAMPLE 5

In $\triangle ABC$, $\angle B = 70°$, $a = 5$, $b = 8$. Determine $\angle A$, correct to one decimal place.

We do not have to worry about c or C here.

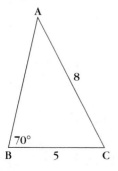

$$\frac{a}{\sin A} = \frac{b}{\sin B}$$

$$\frac{5}{\sin A} = \frac{8}{\sin 70°}$$

$$5\sin 70° = 8\sin A$$

$$\sin A = \frac{5\sin 70°}{8}$$

$$= \frac{5(0.9396926)}{8}$$

$$= 0.5873078$$

$$\therefore \angle A = 36.0°$$

This time, we were given two sides. So, why not use the Cosine Law? Well, the problem is we have the "wrong" two sides or, put another way, we were given the "wrong" angle. If you substitute the given information into the Cosine Law formula, you will readily see that there are two unknowns. This makes it unsolvable.

The conditions for choosing between these laws is summarized below:

GIVEN	COSINE LAW	SINE LAW
3 sides only	yes	no
2 sides, 1 angle	if the given angle is between the given sides	if the given angle is not between the sides
1 side, 2 angles	no	yes

Of course, situations do arise where both laws are useful.

EXAMPLE 6

Solve $\triangle ABC$, given $\angle A = 72°$, $b = 17$, $c = 14$.

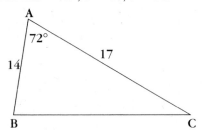

Using the Cosine Law,
$$a^2 = b^2 + c^2 - 2bc\cos A$$
$$= (17)^2 + (14)^2 - 2(17)(14)\cos72°$$
$$= 337.90791$$
$$\therefore a = 18.4$$

At this stage, we could use the Cosine Law again to find $\angle B$ or C. However, it is easier to apply the Sine Law.

$$\frac{a}{\sin A} = \frac{b}{\sin B}$$

$$\frac{18.4}{\sin 72°} = \frac{17}{\sin B}$$

$$\sin B = \frac{17 \sin 72°}{18.4}$$

$$= \frac{17(0.9510565)}{18.4}$$

$$= 0.8786935$$

$$\therefore \angle B = 61.5°$$

Using SATT, $\angle C = 180° - (72° + 61.5°)$

$$= 46.5°$$

There are situations when we must be careful when solving oblique triangles.

EXAMPLE 7
Solve $\triangle BOP$, given $\angle P = 40°$, $b = 11$, $o = 7$.

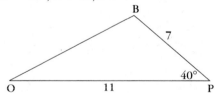

Using the Cosine Law,

$$p^2 = b^2 + o^2 - 2bo\cos P$$
$$= (11)^2 + (7)^2 - 2(11)(7)\cos 40°$$
$$= 52.029156$$
$$\therefore p = 7.2$$

If we use the Sine Law to find $\angle B$,

$$\frac{b}{\sin B} = \frac{p}{\sin P}$$

$$\frac{11}{\sin B} = \frac{7.2}{\sin 40°}$$

$$\sin B = \frac{11 \sin 40°}{7.2}$$

$$= 0.9820366$$

From this we get $\angle B = 79.1°$.

41

However, our diagram indicates $\angle B$ should be obtuse, not acute! The discrepancy is related to the limit of the $\sin^{-1}$ function. This will be described in detail in Chapter 9.

In this case, we should either: (a) use the Cosine Law to find $\angle B$ and complete our solution.

or (b) use the Sine Law to find $\angle O$ (since it is acute) and complete our solution.

For comparison's sake, let's consider both of these routes.

(a) Using the Cosine Law, $b^2 = o^2 + p^2 - 2op\cos B$

$$\therefore \cos B = \frac{b^2 - o^2 - p^2}{-2op}$$

$$= \frac{(11)^2 - (7)^2 - (7.2)^2}{-2(7)(7.2)}$$

$$= \frac{20.16}{-100.8}$$

$$= -0.2$$

$$\therefore \angle B = 101.5°$$

$$\angle O = 180° - (40° + 101.5°)$$

$$= 38.5°$$

(b) Using the Sine Law, $\dfrac{o}{\sin O} = \dfrac{p}{\sin P}$

$$\frac{7}{\sin O} = \frac{7.2}{\sin 40°}$$

$$\sin O = \frac{7\sin 40°}{7.2}$$

$$= \frac{7(0.6427876)}{7.2}$$

$$= 0.6249324$$

$$\therefore \angle O = 38.7°$$

$$\angle B = 180° - (40° + 38.7°)$$

$$= 101.3°$$

As we can see, there are slight differences in the answers we get with these two sets of solutions. This difference is acceptable and is caused by the rounding off of our value for p.

As shown by Example 7, when an angle is suspected to be obtuse, it is wise to avoid using the Sine Law for that particular angle.

PRACTICE EXERCISES

1. Find the measure of the missing side indicated:
    a) For $\triangle KMN$, given $k = 9.2$, $m = 11.8$, and $\angle N = 125°$, find n.
    b) For $\triangle PIG$, given $\angle P = 28°$, $\angle G = 71°$, and $p = 23.1$, find i.

2. Find the measure of the angle indicated:
    a) For $\triangle MAN$, given $m = 49.1$, $a = 38.2$, and $n = 54$, find $\angle A$.
    b) For $\triangle TOY$, given $\angle O = 105°$, $t = 8.1$, and $o = 19.3$, find $\angle T$.

3. Solve each of the following triangles:
    a) $\triangle DEF$, given $d = 10$, $f = 12$, $\angle E = 36.3°$
    b) $\triangle JKL$, given $\angle K = 47.4°$, $\angle L = 72.7°$, $k = 57$
    c) $\triangle CAN$, given $\angle C = 15°$, $a = 21.5$, $n = 12.6$

CHAPTER SIX

Applications involving oblique triangles

The Cosine and Sine Laws are useful in solving many practical problems involving oblique triangles.

EXAMPLE 1

During a golf game, Tom, starting at point T, wants to get the ball to point F before shooting the ball towards the hole at H. If $\angle T = 29°$, $TH = 443.0\ m$, $TF = 328$ m, what is the distance from F to H?

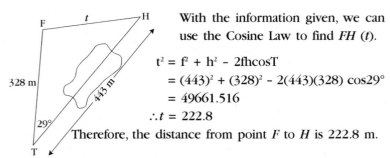

With the information given, we can use the Cosine Law to find FH (t).

$$t^2 = f^2 + h^2 - 2fh\cos T$$
$$= (443)^2 + (328)^2 - 2(443)(328)\cos 29°$$
$$= 49661.516$$
$$\therefore t = 222.8$$

Therefore, the distance from point F to H is 222.8 m.

EXAMPLE 2

Fire towers A and B are 9.8 km apart. A fire at point C is observed from both towers. From tower A, Andy measures $\angle CAB$ as $58°$ while, from tower B, Barbara measures $\angle CBA$ as $48°$. Find the distance from each tower to the fire.

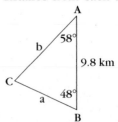

After sketching and labelling our diagram here, we should define the variables.

a = the distance from tower B to the fire
b = the distance from tower A to the fire

Using SATT, $\angle C = 180° - (58° + 48°)$
$$= 74°$$

44

Applying the Sine Law,

$$\frac{a}{\sin 58°} = \frac{9.8}{\sin 74°} \qquad\qquad \frac{b}{\sin 48°} = \frac{9.8}{\sin 74°}$$

$$\therefore a = \frac{9.8 \sin 58°}{\sin 74°} \qquad\qquad \therefore b = \frac{9.8 \sin 48°}{\sin 74°}$$

$$= 8.6 \qquad\qquad\qquad\qquad = 7.6$$

Therefore, tower A is 7.6 km from the fire and tower B is 8.6 km from the fire.

Often, there are times when the Cosine or Sine Law may be used along with the methods for solving right triangles (Chapter 4) to work out a problem.

EXAMPLE 3

Two tracking stations, 18 km apart, measure the angles of elevation of a satellite which is flying west of both stations. The angles are found, to be 43° and 75.6° respectively. Find the altitude of the satellite, ignoring the Earth's curvature.

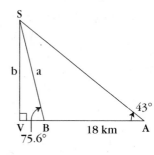

After a diagram has been drawn and labelled, we develop our strategy by working backwards.

ΔSVB is a right triangle, but with no known sides. However, using the information in ΔSBA we can find a and then, determine the height, b, using a trig ratio.

To find a using the Sine Law, we need to determine $\angle BSA$.

$$\angle SBA = 180° - 75.6°$$
$$= 104.4°$$
$$\therefore \angle BSA = 180° - (104.4° + 43°) = 32.6°$$

Using the Sine Law, $\qquad \dfrac{a}{\sin A} = \dfrac{s}{\sin S}$

$$\frac{a}{\sin 43°} = \frac{18}{\sin 32.6°}$$

$$\therefore a = \frac{18 \sin 43°}{\sin 32.6°}$$

$$= 22.785145$$

In $\triangle SVB$, $\frac{b}{a} = \sin 75.6°$

$\therefore\ b = a\sin 75.6°$

$\qquad = 22.785145\ (0.9685831)$

$\qquad = 22.1$

Therefore, the altitude of the satellite is 22.1 km.

EXAMPLE 4

An old apartment building is situated on the bank of a river. From a point directly opposite the building, on the other side of the river, Jody measures an angle of elevation to the top of the building as 14.0°. A line to the base of the building and the baseline to another survey point 200 m downshore, forms an angle of 84.7°. Jody measures an angle of 61.2° between the baseline and a line drawn from this second point to the base of the building. How tall is the building?

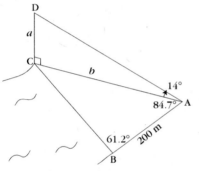

As in Example 3, we don't have enough information to calculate a in $\triangle ACD$. Nonetheless, we can find b in $\triangle ABC$ using the Sine Law and then, use this to calculate a.

$$\angle ACB = 180° - (84.7° + 61.2°)$$

$$= 34.1°$$

$$\frac{b}{\sin B} = \frac{c}{\sin C}$$

$$\frac{b}{\sin 61.2°} = \frac{200}{\sin 34.1°}$$

$$\therefore b = \frac{200\sin 61.2°}{\sin 34.1°}$$

$$= 312.60996$$

$$\frac{a}{b} = \tan 14°$$

Therefore, the apartment building is 77.9 m tall.

$$\therefore a = 312.60996(0.249328)$$

$$= 77.9$$

46

PRACTICE EXERCISES

1. A tree 15 m tall is leaning at an angle of 10° with the vertical. To stop the tree from falling over, a 30 m rope is attached to the top of the tree and fastened to the ground a distance away. What angle does the rope make with the ground?

2. Two highways diverge at 53.2°. If two motorcyclists, starting at the point of divergence, simultaneously take separate routes - one at 55 km/h, the other at 70 km/h - how far apart will they be after 2 h?

3. A radio tower stands erect on top of a hill that is inclined at a 16° angle. If the tower casts a shadow 43.7 m long on the hill, and the angle of elevation of the sun is 46°, find the height of the tower.

4. A hot air balloon is rising vertically from its launch site. From the balloon, Linda sees a duck pond at an angle of depression of 31°. When the balloon rises another 50 m, the angle of depression to the pond is 43°. How far is the pond from the balloon's launch site?

5. To measure the altitude of an airplane, Jake and Janet measure angles of elevation and baseline angles as indicated in the diagram below. What was the altitude of the plane as determined by each of them?

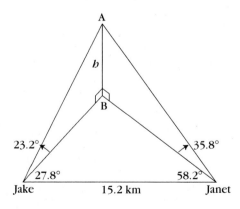

CHAPTER SEVEN

Radian measure

As well as using measurement in degrees, angles can be measured in units called radians. The radian is more directly related to the length of a circle's radius. As a result, it is often more useful for describing angles than the measure in degrees.

An angle of one radian is subtended from the center of a circle by an arc equal to the radius. This is illustrated below.

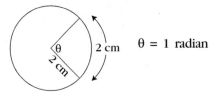

$\theta = 1$ radian

The radian measurement of an angle is given by the following formula:

$$\# \text{ radians} = \frac{\text{arc length}}{\text{radius}}$$

Since the circumference of a circle is equal to $2\pi r$ where r is the radius, then the angle of a complete rotation is:

$$\frac{2\pi r}{r} = 2\pi \text{ radians}$$

But, in degrees, one complete rotation is equal to $360°$.

$$\therefore 360° = 2\pi \text{ rad.}$$

or $\boxed{180° = \pi}$

The symbol for radians can be left out but, the symbol for degrees must be written.

To convert angles measured in degrees to radians, multiply by

$$\frac{\pi}{180}$$

EXAMPLE 1

Convert to radians, expressing the answer in reduced fractional form:

a) $1°$

$$1\left(\frac{\pi}{180}\right) = \frac{\pi}{180} \text{ rad}$$

b) $90°$

$$90\left(\frac{\pi}{180}\right) = \frac{\pi}{2} \text{ rad}$$

EXAMPLE 2

Convert to degrees:

a) $\dfrac{\pi}{3}$

a) $\dfrac{\pi}{3}\left(\dfrac{180}{\pi}\right) = 60°$

b) 5

b) $5\left(\dfrac{180}{\pi}\right) = \dfrac{900}{\pi}$

$$= 286.48°$$

You should try to become familiar with the trig ratios of the special triangles and their related angles using radian units as well as degree units. Refer back to Chapter 3 if you need to review the work on special triangles.

The following diagram may be helpful in studying these important conversions.

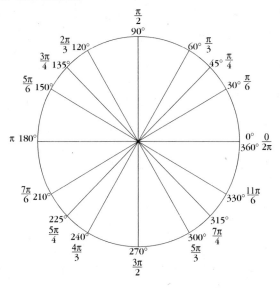

EXAMPLE 3

Evaluate each of the following without using a calculator:

a) $\sin\dfrac{\pi}{6}$

b) $\cos\dfrac{5\pi}{4}$

a) $\sin\dfrac{\pi}{6} = \sin 30°$

$\qquad = \dfrac{1}{2}$

b) $\cos\dfrac{5\pi}{4} = \cos\left(\pi + \dfrac{\pi}{4}\right)$

$\qquad = \cos(180° + 45°)$

$\qquad = -\cos 45°$

$\qquad = -\dfrac{1}{\sqrt{2}}$

Note that the above questions were done by converting the radians to degrees. Ideally, this conversion should not be necessary. It would be preferable if you could "think" and evaluate in radians, rather than converting to degrees first. This is similar to speaking two or more languages. It is more efficient if you can think in the language you are speaking, rather than translating in your head.

 Just as a calculator can be used to evaluate trig ratios for angles measured in degrees, so it can be used for radian measurements. There are two important operations to be aware of:

- your calculator must be in RAD mode before you start.
- the π button should be used where appropriate.

EXAMPLE 4

a) Evaluate, rounding to four decimal places:

i) $\sin\dfrac{\pi}{8}$

ii) $\tan(-3)$

i) $\sin\dfrac{\pi}{8} = \sin 0.392699$

$\qquad = 0.3827$

ii) $\tan(-3) = 0.1425$

b) Determine the value of the angle, rounded to two decimal places:

i) $\cos\theta = -0.1328$ ii) $\csc^{-1}3.9$

i) $\theta = 1.70$ ii) $\csc^{-1}3.9 = \sin^{-1}\dfrac{1}{3.9}$

$$= \sin^{-1}0.2564$$
$$= 0.26$$

PRACTICE EXERCISES

1. Convert to radians, expressing answers in reduced fractional form:

a) $5°$ b) $36°$ c) $-350°$ d) $270°$ e) $20°$

2. Convert to degrees:

a) $\dfrac{\pi}{8}$ b) $\dfrac{3\pi}{5}$ c) -3π d) $\dfrac{\pi}{12}$ e) $-\dfrac{5\pi}{2}$

3. Draw the two special triangles, labelling the angles in radians.

4. Evaluate without using a calculator:

a) $\sin\dfrac{\pi}{4}$ b) $\cos\dfrac{\pi}{3}$ c) $\tan\dfrac{\pi}{6}$

d) $\sin\dfrac{\pi}{3}$ e) $\tan\dfrac{3\pi}{4}$ f) $\cos\dfrac{2\pi}{3}$

g) $\sin\dfrac{5\pi}{6}$ h) $\cos\dfrac{4\pi}{3}$ i) $\tan\dfrac{11\pi}{6}$

j) $\cos\dfrac{7\pi}{6}$ k) $\sin\dfrac{5\pi}{4}$ l) $\tan\left(-\dfrac{7\pi}{4}\right)$

5. Determine the value of the following trig ratios, rounded to four decimal places:

a) $\cos\dfrac{\pi}{12}$ b) $\tan\dfrac{2\pi}{5}$ c) $\sin 3.8$ d) $\cos(-2)$ e) $\sec 0.5$

6. Evaluate, rounded to four decimal places:

a) $\sin^{-1}0.4887$ b) $\cos^{-1}0.7071068$ c) $\tan^{-1}1$

d) $\sin^{-1}(-0.6925)$ e) $\cot^{-1}2.0711$

CHAPTER EIGHT

Graphs of trigonometric functions and their inverses

Using tables of values of trig ratios and their corresponding angles, we can draw graphs of the trig functions. However, before we consider these, there are trig ratios for some key angles which we have not yet discussed. These angles include 0°, 90°, 180°, 270° and their coterminal angles.

Starting at the initial arm and rotating the terminal arm around to the positions on the axes, we can form any of these angles. Let's consider points on a unit circle, $x^2 + y^2 = 1$, at these positions.

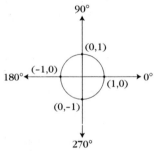

Using the same definitions for the trig ratios, developed in Chapter 4, with $x, y,$ and r, we get the following results:

sin 0° = 0	sin 90° = 1
cos 0° = 1	cos 90° = 0
tan 0° = 0	tan 90° = undefined
sin 180° = 0	sin 270° = -1
cos 180° = -1	cos 270° = 0
tan 180° = 0	tan 270° = undefined

Just as we saw in Chapter 2 on similar triangles, the value of the radius doesn't change the trig ratios.

Using the information in the above box, in addition to our work based on the special triangles, we can develop a table of values for each of the trig functions. Because of their association with circles, it is customary to express the angles in radians, although it is also acceptable to use degrees. Because of this association, the trig functions are often referred to as circular functions.

Let's start with $y = \sin\theta$ where y represents the trig ratio.

θ	0	$\frac{\pi}{6}$	$\frac{\pi}{4}$	$\frac{\pi}{3}$	$\frac{\pi}{2}$	$\frac{2\pi}{3}$	$\frac{3\pi}{4}$	$\frac{5\pi}{6}$	π	$\frac{7\pi}{6}$	$\frac{5\pi}{4}$	$\frac{4\pi}{3}$	$\frac{3\pi}{2}$	$\frac{5\pi}{3}$	$\frac{7\pi}{4}$	$\frac{11\pi}{6}$	2π
$\sin\theta$	0	$\frac{1}{2}$	$\frac{1}{\sqrt{2}}$	$\frac{\sqrt{3}}{2}$	1	$\frac{\sqrt{3}}{2}$	$\frac{1}{\sqrt{2}}$	$\frac{1}{2}$	0	$-\frac{1}{2}$	$-\frac{1}{\sqrt{2}}$	$-\frac{\sqrt{3}}{2}$	-1	$-\frac{\sqrt{3}}{2}$	$-\frac{1}{\sqrt{2}}$	$-\frac{1}{2}$	0

The following steps, illustrated for $y = \sin\theta$, can be used to graph any of the basic trig functions.

1. Label the axes as shown. Divide each horizontal distance of radians into 12 equal sections. (Thus, each 2 sections represents $\frac{\pi}{4}$ units; each 3 sections represents $\frac{\pi}{4}$ units). Let $y = 1$ be about $\frac{1}{3}$ as long as π.

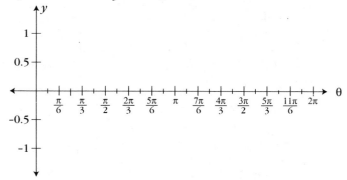

2. Plot the points where $y = 0$, 1, and -1.

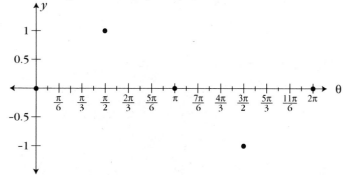

3. Plot the other points. $\left(\dfrac{\sqrt{3}}{2} \approx 0.9, \ \dfrac{1}{\sqrt{2}} \approx 0.7 \right)$

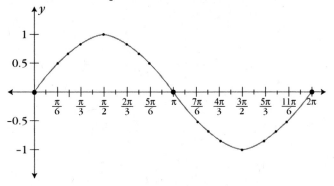

4. Connect all the points to form a smooth curve.

This gives us one complete cycle of $y = \sin\theta$ and, graphically, represents the sine values for all angles from 0 to 2π.

Once you are familiar with the shape of this function, you can sketch the graph more quickly by:

1. dividing the horizontal axis into multiples of $\dfrac{\pi}{2}$
2. plotting the points where $y = 0, 1, -1$
3. connecting the points to form a smooth curve.

As we have seen, angles need not be limited to these values. We can use coterminal as well as negative angles, if we wish. The curve can be extended indefinitely in both directions. The following sketch represents $y = \sin\theta$ where $-3\pi \leq \theta \leq 3\pi$.

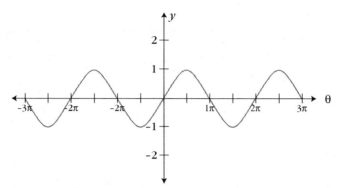

Because the values of y repeat at equal intervals of θ, $y = \sin\theta$ and the other trig functions are called **periodic functions**.

To graph $y = \cos\theta$, we'll use a similar table of values.

θ	0	$\frac{\pi}{6}$	$\frac{\pi}{4}$	$\frac{\pi}{3}$	$\frac{\pi}{2}$	$\frac{2\pi}{3}$	$\frac{3\pi}{4}$	$\frac{5\pi}{6}$	π	$\frac{7\pi}{6}$	$\frac{5\pi}{4}$	$\frac{4\pi}{3}$	$\frac{3\pi}{2}$	$\frac{5\pi}{3}$	$\frac{7\pi}{4}$	$\frac{11\pi}{6}$	2π
$\cos\theta$	1	$\frac{\sqrt{3}}{2}$	$\frac{1}{\sqrt{2}}$	$\frac{1}{2}$	0	$-\frac{1}{2}$	$-\frac{1}{\sqrt{2}}$	$-\frac{\sqrt{3}}{2}$	-1	$-\frac{\sqrt{3}}{2}$	$-\frac{1}{\sqrt{2}}$	$-\frac{1}{2}$	0	$\frac{1}{2}$	$\frac{1}{\sqrt{2}}$	$\frac{\sqrt{3}}{2}$	1

Using the same steps to draw the graph as was done for $y = \sin\theta$, we get:

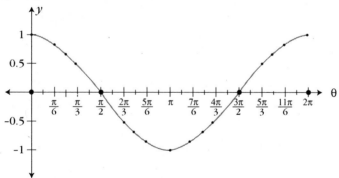

The shape of $y = \cos\theta$ is like $y = \sin\theta$ except the wave starts at $(0, 1)$ instead of $(0, 0)$.

The table of values for $y = \tan\theta$ is completed below:

θ	0	$\frac{\pi}{6}$	$\frac{\pi}{4}$	$\frac{\pi}{3}$	$\frac{\pi}{2}$	$\frac{2\pi}{3}$	$\frac{3\pi}{4}$	$\frac{5\pi}{6}$	π	$\frac{7\pi}{6}$	$\frac{5\pi}{4}$	$\frac{4\pi}{3}$	$\frac{3\pi}{2}$	$\frac{5\pi}{3}$	$\frac{7\pi}{4}$	$\frac{11\pi}{6}$	2π
$\tan\theta$	1	$\frac{1}{\sqrt{3}}$	1	$\sqrt{3}$	$-$	$-\sqrt{3}$	-1	$-\frac{1}{\sqrt{3}}$	0	$\frac{1}{\sqrt{3}}$	1	$\sqrt{3}$	$-$	$-\sqrt{3}$	-1	$-\frac{1}{\sqrt{3}}$	0

Note that tanθ is undefined for $\theta = \frac{\pi}{2}, \frac{3\pi}{2}$ At these points, the graph has vertical asymptotes. These points are half-way between the points where $y = 0$.

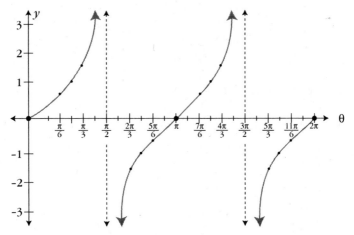

As you may recall, a function, f, can have an inverse function, f^{-1}, which is determined by interchanging x and y.

For instance: $y = \sin x$ is a function.
$x = \sin y$ which, by convention, is rewritten as $y = \sin^{-1} x$, is the inverse function.

However, unless a function has a different value of y for each value of the domain (x or θ), then its inverse is not a function.

In order for the inverses of the trig functions to be functions, mathematicians have decided to restrict the x values of the primary trig functions as follows:

$$y = \sin x, -\frac{\pi}{2} \le x \le \frac{\pi}{2}$$
$$y = \cos x, 0 \le x \le \pi$$
$$y = \tan x, -\frac{\pi}{2} \le x \le \frac{\pi}{2}.$$

In Chapter 4, the boundaries for using the inverse trig functions on the calculator were mentioned. These restrictions are due to this condition for inverse functions. As a result of these restrictions, the graphs of the inverse trig functions are not periodic. They have a definite beginning and end.

Remembering the interchanging of x and y, a table of values for $y = \sin^{-1}x$ is shown below.

$\sin^{-1}x$	-1	$-\dfrac{\sqrt{3}}{2}$	$-\dfrac{1}{\sqrt{2}}$	$-\dfrac{1}{2}$	0	$\dfrac{1}{2}$	$\dfrac{1}{\sqrt{2}}$	$\dfrac{\sqrt{3}}{2}$	1
y	$-\dfrac{\pi}{2}$	$-\dfrac{\pi}{3}$	$-\dfrac{\pi}{4}$	$-\dfrac{\pi}{6}$	0	$\dfrac{\pi}{6}$	$\dfrac{\pi}{4}$	$\dfrac{\pi}{3}$	$\dfrac{\pi}{2}$

PRACTICE EXERCISES

1. Draw the graph of $y = \cos\theta$ on the interval $-2\pi \le \theta \le 2\pi$.

2. Using a table of values for $0 \le \theta \le 2\pi$, draw a graph for each of the reciprocal trig functions.

3. Draw a graph for the inverse function $y = \cos^{-1}x$.

CHAPTER NINE

Transformations and applications of sine and cosine graphs

Like other types of functions, transformations can be performed on the basic trig graphs. Transformations of the sine and cosine graphs have various applications in science. So we will focus on transformations of these functions.

Because of their scientific associations and wave-like shape, special terms are used to describe the features of these graphs.

Consider the sketch of $y = \sin\theta$ below.

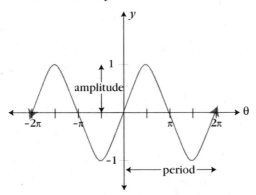

amplitude = the height or depth of a wave

period   = the interval in the domain for which one wave cycle is completed

For $y = \sin\theta$ and $y = \cos\theta$, the amplitude is 1 and the period is 2π.

Now consider the following graph.

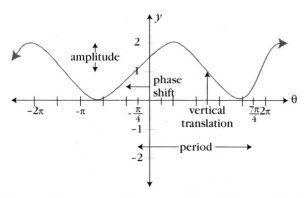

phase shift = the amount by which the basic graph is shifted left or right.

(This is a specialized name for a horizontal translation.)

vertical translation = the amount by which the basic graph is shifted up or down.

In the graph above, the basic sine function has been shifted left $\frac{\pi}{4}$ and translated up 1.

The general equations for sine or cosine graphs can be expressed as follows:

$$y = a\sin k\,(\theta + c) + d$$
$$y = a\cos k\,(\theta + c) + d$$
a = amplitude
$\quad = \frac{2\pi}{k}$ period
c = phase shift
d = vertical translation

You might find these steps useful for sketching transformed trig functions:

1. Divide the horizontal scale as suggested in Chapter 8 so each block of 2π units has 12 parts.
2. Sketch the basic trig graph over the domain required.
3. On the same set of axes, (using a different color) use the phase shift to find a starting point for the new graph and use the period to mark the end of one wave.

4. Divide this interval into 4 equal parts. Use the amplitude to mark the high and low points. Mark the ends of the other parts as zero points.
5. Join the points to form one wave. Extend/shorten this wave pattern according to the domain required.
6. Move all points according to the vertical translation (using a different color), if applicable, and join these points to complete the sketch.

Once you become familiar with the above process, you may feel comfortable about cutting corners. Nonetheless, do so with caution!

EXAMPLE 1

For the function $y = 3\sin2\theta$: a) state the amplitude, period, phase shift and vertical translation

b) sketch the graph for $0 \le \theta \le 2\pi$.

a) amplitude = 3
period = $\frac{2\pi}{2} = \pi$
phase shift = none
vertical translation = none

b) Steps 1 to 3 have been completed below. The basic sketch of $y = \sin\theta$ for $0 \le \theta \le 2\pi$ has been drawn. The beginning and end of one period for $y = 3\sin 2\theta$ have been marked.

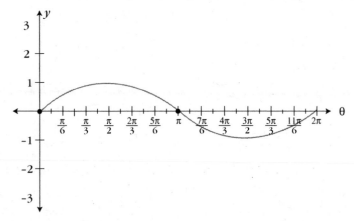

Using step 4, we divide the period by 4 to give us an interval of $\frac{\pi}{4}$. Incorporating the new amplitude of 3 to mark the high and low points, and marking points where $y = 0$, leads us to the following:

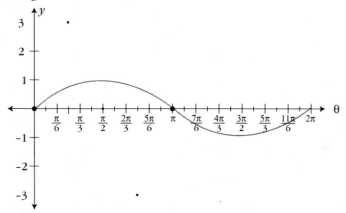

Joining these points and extending the pattern for the domain required, gives us the final sketch. As usual, the curve is labelled.

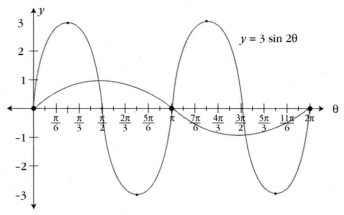

$y = 3 \sin 2\theta$

EXAMPLE 2

Sketch the graph of $y = \cos 3(x + \frac{\pi}{4}) + 1$, $-\pi \leq x \leq \pi$.

Although not asked for, it is wise to start by writing the special features of the function. This helps us to organize the sketch.

amplitude = 1
period = $\frac{2\pi}{3}$
phase shift = $\frac{\pi}{4}$ left
vertical translation = 1 up

As in Example 1, we complete steps 1 to 3 in the first stage below.

The graph of $y = \cos x$ is sketched on the required domain. Due to the phase shift, the transformed function starts at $-\frac{\pi}{4}$. To find the end of the period, we count right $\frac{8\pi}{12} = \left(= \frac{2\pi}{3} \right)$ units.

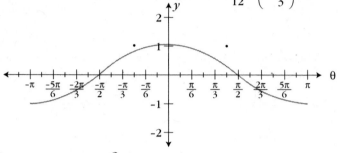

Applying step 4: $\frac{2\pi}{3} \div 4 = \frac{\pi}{6}$.

Using this to mark the low and zero points, and performing step 5 to extend the pattern for the domain, $-\pi \leq x \leq \pi$, we get the following graph:

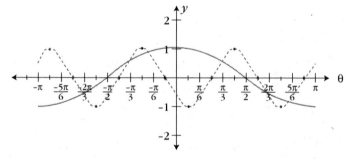

Now, all the points are moved up and the curve is redrawn to get the final sketch.

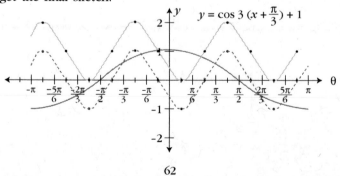

$$y = \cos 3 \left(x + \frac{\pi}{3} \right) + 1$$

EXAMPLE 3

Sketch the graph of $y = 2\sin(4\theta - \frac{\pi}{3}) - 1$ where $-\frac{\pi}{2} \le \theta \le \frac{\pi}{2}$.

This equation is not in the $y = a\sin k(\theta + c) + d$ form. We must factor k (i.e. 4) first.

$y = 2\sin(4\theta - \frac{\pi}{3}) - 1$

$y = 2\sin 4(\theta - \frac{\pi}{12}) - 1$

amplitude = 2

period $= \frac{2\pi}{4} = \frac{\pi}{2}$

phase shift $= \frac{\pi}{12}$ right

vertical translation = 1 down

Steps 1 and 2: The graph of $y = \sin\theta$ where $-\frac{\pi}{2} \le \theta \le \frac{\pi}{2}$ is drawn.

Step 3: Using the phase shift, $\frac{\pi}{12}$ right, we locate the new starting point. The period, $\frac{6\pi}{12}$ ($= \frac{\pi}{2}$) units leads us to the end point of the wave. Since $\frac{7\pi}{12}$ is not in the required domain, this last portion of the curve is left out of the final sketch. Nonetheless, it serves as a guide for steps 4 and 5.

Step 4: $\dfrac{\pi}{2} \div 4 = \dfrac{\pi}{8} \left(\dfrac{\pi}{8} = 1.5 \times \dfrac{\pi}{12} \right)$

The high and low points are marked with an amplitude of 2. The zero points are marked.

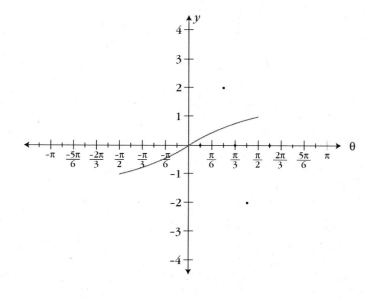

Step 5: We extend the pattern over the domain required, omitting the part beyond.

Step 6: Shift all points down 1 and connect the points to complete the sketch.

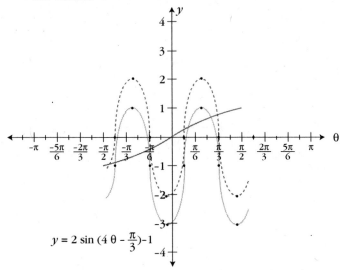

$$y = 2 \sin (4\,\theta - \frac{\pi}{3}) - 1$$

Transformations of the sine or cosine functions can be used to express wave-like phenomena or the circular motion of various objects.

EXAMPLE 4

A ferris wheel has a radius of 6 m and rotates once every 10 seconds. If the bottom of the wheel is 1 m above the ground:

a) write an equation to represent a person's height above the ground starting at the top of the wheel.

b) draw a graph showing one cycle.

a) amplitude = radius = 6 m

period = time for 1 rotation = 10 s

$\therefore \frac{2\pi}{k} = 10$

$\therefore k = \frac{2\pi}{10} = \frac{\pi}{5}$

The center of the ferris wheel is 1 + 6 = 7 m above the ground.

$\therefore$ vertical translation = 7 m up

Since the person starts at the top, we can use the cosine function to express height, y, in terms of time, t:

$\therefore y = 6 \cos \frac{\pi}{5} t + 7$ is the required equation.

b) Because the radius of the ferris wheel is 6 m and the center is 7 m above the ground, the greatest height is 13 m and the lowest is 1 m. In addition to the other information we are given, the graph for one rotation can be drawn as follows:

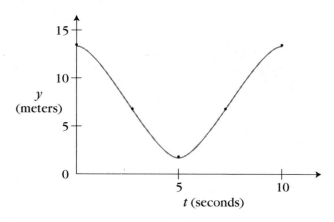

PRACTICE EXERCISES

1. Sketch each graph on the interval indicated:

   a) $y = 2\sin 4\theta, -\pi \le \theta \le \pi$

   b) $y = 3\cos 2\left(x - \dfrac{\pi}{4}\right) + 1, 0 \le x \le 2\pi$

   c) $y = \dfrac{1}{2}\sin\left(3x + \dfrac{\pi}{2}\right) - 2, -\pi \le x \le \pi$

2. During high tide, the water depth in an ocean harbour is 20 m and during low tide, it is 8 m. If the tide cycle is 12 h:

   a) write an expression for the water depth, t hours after high tide.

   b) Sketch a graph of the function for a duration of 24 hours.

3. Susie pedals her bicycle at a constant rate, such that the height of her foot above the ground is given by

$$h = 12\sin 2\pi\left(t - \frac{1}{4}\right) + 17$$

where h is the height of her foot in centimeters and t is time in seconds.

a) What is the height of her foot at $t = 0$ seconds?
b) What is the radius of the pedal arm?
c) How much time does it take for her foot to make one rotation?
d) Sketch a graph to show the height of Susie's foot for two rotations.
e) Write a cosine function that expresses the height of Susie's other foot.

CHAPTER TEN

Solving trigonometric equations

When solving a trig equation, we are generally trying to determine all the angles which make the equation true. Because of the periodic nature of the trig functions, there is usually more than one solution, depending on the domain specified.

There are two types of strategies that can be used:

(i) graphical
   1. sketch the graph of the function represented in the equation.
   2. locate points which satisfy the equation.

(ii) circular
   1. determine the acute angle which gives the desired trig ratio.
   2. starting at the initial arm, rotate the terminal arm on the Cartesian grid through the positive part of the domain.
   3. using the CAST rule, locate the related angles which give the desired trig ratio.
   4. repeat step 3 rotating through the negative part of the domain.

Usually, textbooks only demonstrate the graphical strategy. However, the circular approach is often more accurate. It is not as complicated as it may seem. In fact, it is based on the work we already covered in Chapter 4.

In the following two examples, let's consider solutions using each of these strategies.

EXAMPLE 1

Solve for x: $\sin x = \frac{1}{2}$, $-2\pi \le x \le 2\pi$.

Solution (i): We sketch the graph of $y = \sin x$ on the domain given.

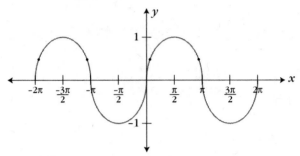

Looking at the graph, we determine

$$y = \frac{1}{2} \text{ when } x = -\frac{11\pi}{6}, -\frac{7\pi}{6}, \frac{\pi}{6}, \frac{5\pi}{6}.$$

Solution (ii): We know $\sin \frac{\pi}{6} = \frac{1}{2}$

$$\therefore x = \frac{\pi}{6}$$

Rotating the terminal arm on the Cartesian grid from 0 to 2π, and using the CAST rule, results in the following diagram and conclusion:

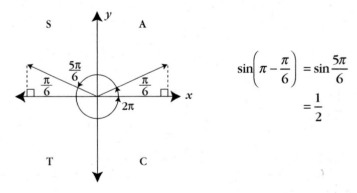

$$\sin\left(\pi - \frac{\pi}{6}\right) = \sin \frac{5\pi}{6}$$

$$= \frac{1}{2}$$

Rotating from 0 to -2π, the ratio for sine is $\frac{1}{2}$ if

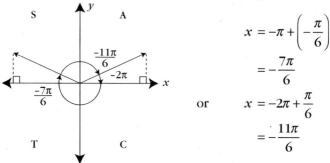

$$x = -\pi + \left(-\frac{\pi}{6}\right)$$

$$= -\frac{7\pi}{6}$$

or

$$x = -2\pi + \frac{\pi}{6}$$

$$= -\frac{11\pi}{6}$$

(Remember when rotating clockwise, we add a negative angle. When rotating counterclockwise, we add a positive angle.

Therefore, $x = -\frac{11\pi}{6}, -\frac{7\pi}{6}, \frac{\pi}{6}, \frac{5\pi}{6}$, is the solution

EXAMPLE 2

Solve for x : $\cos 2x = -\frac{1}{\sqrt{2}}$, $-\frac{\pi}{2} \le x \le \frac{\pi}{2}$

Solution (i):   We sketch the graph, remembering the period is π here.

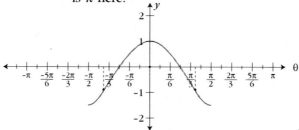

Reading points from the graph where $y = -\frac{1}{\sqrt{2}}$,

we find $x = -\frac{3\pi}{8}, \frac{3\pi}{8}$.

Solution (ii):   In order to solve for x on $-\frac{\pi}{2} \le x \le \frac{\pi}{2}$,

we first solve for $2x$ on $-\pi \le 2x \le \pi$ (multiply by 2)

We know $\cos \frac{\pi}{4} = \frac{1}{\sqrt{2}}$

$$\therefore x = \frac{\pi}{4}$$

but, we need to find related angles for which the

trig ratio is negative $\frac{1}{\sqrt{2}}$.

Rotating the terminal arm from 0 to π,

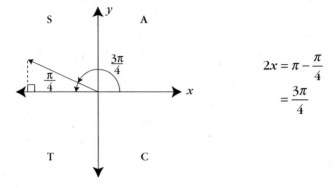

$$2x = \pi - \frac{\pi}{4}$$

$$= \frac{3\pi}{4}$$

Rotating the terminal arm from 0 to $-\pi$,

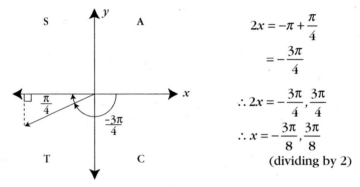

$$2x = -\pi + \frac{\pi}{4}$$

$$= -\frac{3\pi}{4}$$

$$\therefore 2x = -\frac{3\pi}{4}, \frac{3\pi}{4}$$

$$\therefore x = -\frac{3\pi}{8}, \frac{3\pi}{8}$$

(dividing by 2)

Sometimes, as you may have found in the previous example, it is hard to accurately read/interpret the angles from a graph. Hence, you may find the circular strategy more reliable.

For the sake of brevity, we'll use the circular strategy only in the remaining examples.

EXAMPLE 3

Solve for x: $2\sin^2 x - \sin x - 1 = 0, 0 \le x \le 2\pi$

We begin by factoring the equation and setting each factor equal to zero:

$$2\sin^2 x - \sin x - 1 = 0$$

$$(2\sin x + 1)(\sin x - 1) = 0$$

$$2\sin x + 1 = 0 \qquad \text{or} \qquad \sin x - 1 = 0$$

Rearranging, $\sin x = -\frac{1}{2}$

Solving, $a = \frac{\pi}{6}$

but we need related angles for the trig ratio $-\frac{1}{2}$

$\sin x = 1$

$x = \frac{\pi}{2}$

This is the only solution for this equation on the domain given.

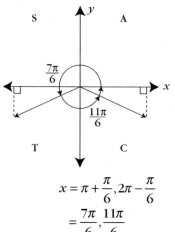

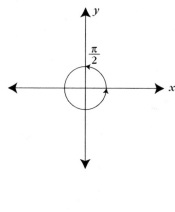

$$x = \pi + \frac{\pi}{6}, 2\pi - \frac{\pi}{6}$$

$$= \frac{7\pi}{6}, \frac{11\pi}{6}$$

Therefore, $x = \frac{\pi}{2}, \frac{7\pi}{6}, \frac{11\pi}{6}$, is the solution.

Sometimes, we are asked to write the general solution for a trig equation. This means we must write all real solutions from negative infinity to positive infinity. Of course, we cannot write an infinite number of solutions. By convention, this type of solution is written as one or more equations using an additional variable such as n.

EXAMPLE 4

Find the general solution for $\sin x = \frac{\sqrt{3}}{2}$

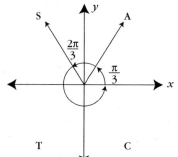

First, we rotate the terminal arm from 0 to 2π and identify the angles which satisfy this equation.

Doing this we get:

$$x = \frac{\pi}{3}, \frac{2\pi}{3}$$

Since the period of y = sinx is 2π, then, by adding or subtracting whole number multiples of 2π to or from either of the above values for x, we obtain coterminal angles which also satisfy the equation.

For example, the following are other solutions for the given equation:

$$\frac{\pi}{3} + 2\pi, \frac{\pi}{3} + 2(2\pi), \frac{\pi}{3} - 2\pi, \frac{\pi}{3} - 2(2\pi)$$

Therefore, the general solution is: $x = \dfrac{\pi}{3} + 2\pi n$

or $x = \dfrac{2\pi}{2} + 2\pi n,$ where $n \in I$

(*I* is the set of integers)

EXAMPLE 5

State the general solution for cos $2x = -\frac{1}{2}$.

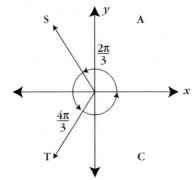

Rotating from 0 to 2π we get

$$2x = \frac{2\pi}{3}, \frac{4\pi}{3}$$

For the general soluton: $2x = \dfrac{2\pi}{3} + 2\pi n$ or $2x = \dfrac{4\pi}{3} + 2\pi n, n \in I$

Dividing by 2: $\therefore x = \dfrac{\pi}{3} + \pi n$ or $x = \dfrac{2\pi}{3} + \pi n, n \in I$

EXAMPLE 6

State the general solution for tan $x = 1$.

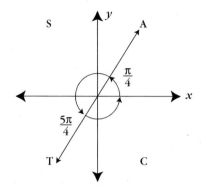

Rotating from 0 to 2π, we get

$$x = \frac{\pi}{4}, \frac{5\pi}{4}$$

Since the difference between these angles is π, we only need one equation to express the general solutions here.

Therefore, the general solution is $x = \frac{\pi}{4} + \pi n$, $n \varepsilon I$

PRACTICE EXERCISES

1. Solve for x on the given interval:

   a) $\sin x = \frac{\sqrt{3}}{2}$, $-\pi \le x \le \pi$

   b) $\tan x = \frac{1}{\sqrt{3}}$, $-2\pi \le x \le 2\pi$

   c) $\cos 2x = 1$, $-\pi \le x \le \pi$

   d) $\left(\cos x - \frac{1}{2}\right)(\tan x + 1) = 0$, $0 \le x \le 2\pi$

   e) $\cos^2 2x + \cos 2x = 0$, $-2\pi \le x \le 2\pi$

2. Determine the general equation:

   a) $\cos x = \frac{1}{\sqrt{2}}$

   b) $\sin 2x = -\frac{1}{2}$

   c) $\tan 2x = \sqrt{3}$

CHAPTER ELEVEN

Trigonometric identities

An identity is a statement of equality involving variables. To prove identities are true, we make use of identities we already know.

In Chapter 2, the reciprocal identities were stated.

Reciprocal identities

$$\csc \theta = \frac{1}{\sin \theta}$$

$$\sec \theta = \frac{1}{\cos \theta}$$

$$\cot \theta = \frac{1}{\tan \theta}$$

Using substitutions with x, y, and r, which were introduced in Chapter 4, we can develop some other basic identities.

$$\frac{\sin \theta}{\cos \theta} = \frac{\frac{y}{r}}{\frac{x}{r}} \qquad\qquad \frac{\cos \theta}{\sin \theta} = \frac{\frac{x}{r}}{\frac{y}{r}}$$

$$= \frac{y}{r} \times \frac{r}{x} \qquad\qquad = \frac{x}{r} \times \frac{r}{y}$$

$$= \frac{y}{x} \qquad\qquad\qquad = \frac{x}{y}$$

$$= \tan \theta \qquad\qquad\quad = \cot \theta$$

The above steps establish the quotient identities.

Quotient identities

$$\tan\theta = \frac{\sin\theta}{\cos\theta}$$

$$\cot\theta = \frac{\cos\theta}{\sin\theta}$$

Expanding on the same theme,

$$\sin^2\theta + \cos^2\theta = \frac{y^2}{r^2} + \frac{x^2}{r^2}$$
$$= \frac{x^2 + y^2}{r^2}$$
$$= \frac{r^2}{r^2}$$
$$= 1$$

$$1 + \tan^2\theta = 1 + \frac{y^2}{x^2}$$
$$= \frac{x^2 + y^2}{x^2}$$
$$= \frac{r^2}{x^2}$$
$$= \sec^2\theta$$

$$1 + \cot^2\theta = 1 + \frac{x^2}{y^2}$$
$$= \frac{y^2 + x^2}{y^2}$$
$$= \frac{r^2}{y^2}$$
$$= \csc^2\theta$$

Thus, we can summarize the Pythagorean identities.

Pythagorean identities

$$\sin^2\theta + \cos^2\theta = 1$$
$$1 + \tan^2\theta = \sec^2\theta$$
$$1 + \cot^2\theta = \csc^2\theta$$

All of the above identities can be used to prove other identities true by using substitution. Doing these proofs can be fun like solving a puzzle. Nonetheless, some of them can be challenging. The steps needed for a proof may not be obvious. However, the following set of suggestions may be helpful.

Suggestions for proving trig identities

1. Write each step clearly. Avoid skipping steps which you think are obvious.
2. Simplify one side at a time, starting with the more complicated side.
3. Express the reciprocal trig functions in terms of $\sin\theta$ and/or $\cos\theta$.
4. Use a Pythagorean identity if it helps simplify the expression.

EXAMPLE 1
Prove $\tan\theta\csc\theta = \sec\theta$

Let's start with the more complicated left side (L.S.).

$$\text{L.S.} = \tan\theta\csc\theta$$

$$= \frac{\sin\theta}{\cos\theta}\cdot\frac{1}{\sin\theta}$$

$$= \frac{1}{\cos\theta}$$

$$= \sec\theta$$

Now, we can write a concluding statement.

$$\therefore \tan\theta\csc\theta = \sec\theta$$

76

Because the right side was simple, we were able to change the left side of the identity and make it equal to the right side. Other times, it is better to change both sides until they are identical.

EXAMPLE 2

Prove $\dfrac{\sec^2 \theta}{1-\cos^2 \theta} = \csc^2 \theta + \sec^2 \theta$

Here both sides are somewhat complex. So, we will adjust both of them.

We can start by substituting the reciprocal trig functions with functions of sine and cosine.

We can also simplify the L.S. using a Pythagorean identity.

$\therefore \sin^2 \theta + \cos^2 \theta = 1$

$\sin^2 \theta = 1 - \cos^2 \theta$

$$\text{L.S.} = \dfrac{\sec^2 \theta}{1-\cos^2 \theta} \qquad\qquad \text{R.S.} = \csc^2 \theta + \sec^2 \theta$$

$$= \dfrac{1}{\cos^2\theta} \cdot \dfrac{1}{\sin^2 \theta} \qquad\qquad = \dfrac{1}{\sin^2 \theta} + \dfrac{1}{\cos^2 \theta}$$

$$= \dfrac{1}{\cos^2 \theta \sin^2 \theta} \qquad\qquad = \dfrac{1}{\cos^2 \theta \sin^2 \theta}$$

$$\therefore \dfrac{\sec^2 \theta}{1-\cos^2 \theta} = \csc^2 \theta + \sec^2 \theta$$

Often, it is necessary to expand brackets or use factoring within a proof, as shown in the following example.

EXAMPLE 3

Prove $(1 + \tan x)^2 = \sec^2 x + 2\tan x$

$$\text{L.S.} = (1 + \tan x)^2$$
$$= 1 + 2\tan x + \tan^2 x$$
$$= 1 + \tan^2 x + 2\tan x$$
$$= \sec^2 x + 2\tan x$$
$$\therefore (1 + \tan x)^2 = \sec^2 x + 2\tan x$$

Note that, after expanding the bracket, we had $2 \tan x$ which is also part of the R.S. Hence, we could leave it and concentrate on reorganizing the rest of the L.S. in order to complete the proof.

Referring back again to Chapter 4, we developed several formulas, in relation to the CAST rule, which enable us to evaluate trig ratios for angles related to the special triangles. We can use these to prove other identities!

EXAMPLE 4

Prove $\cos(-\theta) + \cos(180° - \theta) = \cos(180° + \theta) + \cos\theta$

$$\text{L.S.} = \cos(-\theta) + \cos(180° - \theta) \qquad \text{R.S.} = \cos(180° + \theta) + \cos\theta$$
$$= \cos\theta + (-\cos\theta) \qquad\qquad\qquad = -\cos\theta + \cos\theta$$
$$= 0 \qquad\qquad\qquad\qquad\qquad\qquad = 0$$
$$\therefore \cos(-\theta) + \cos(180° - \theta) = \cos(180° + \theta) + \cos\theta$$

The addition and subtraction formulas are other important identities encountered in trigonometry. To demonstrate these, consider the following example.

EXAMPLE 5

Prove $\cos\left(\dfrac{\pi}{3} + \dfrac{\pi}{6}\right) = \cos\dfrac{\pi}{3}\cos\dfrac{\pi}{6} - \sin\dfrac{\pi}{3}\sin\dfrac{\pi}{6}$

$$\text{L.S.} = \cos\left(\frac{\pi}{3} + \frac{\pi}{6}\right) \qquad\qquad \text{R.S.} = \cos\frac{\pi}{3}\cos\frac{\pi}{6} - \sin\frac{\pi}{3}\sin\frac{\pi}{6}$$

$$= \cos\left(\frac{2\pi}{6} + \frac{\pi}{6}\right) \qquad\qquad\qquad = \left(\frac{1}{2}\right)\left(\frac{\sqrt{3}}{2}\right) - \left(\frac{\sqrt{3}}{2}\right)\left(\frac{1}{2}\right)$$

$$= \cos\frac{\pi}{2} \qquad\qquad\qquad\qquad\qquad = \frac{\sqrt{3}}{4} - \frac{\sqrt{3}}{4}$$

$$= 0 \qquad\qquad\qquad\qquad\qquad\qquad = 0$$

$$\therefore \cos\left(\frac{\pi}{3} + \frac{\pi}{6}\right) = \cos\frac{\pi}{3}\cos\frac{\pi}{6} - \sin\frac{\pi}{3}\sin\frac{\pi}{6}$$

As it turns out, the identity works for any pair of angles. So, it can be written generally as:

$$\cos(x + y) = \cos x \cos y - \sin x \sin y$$

You should be aware that
$$\cos(x + y) \neq \cos x + \cos y$$

This can be demonstrated using the angles in the previous example.

As was shown $\cos\left(\dfrac{\pi}{3} + \dfrac{\pi}{6}\right) = 0$

but $\quad \cos\dfrac{\pi}{3} + \cos\dfrac{\pi}{6} = \dfrac{1}{2} + \dfrac{\sqrt{3}}{2} = \dfrac{1+\sqrt{3}}{2}$

$\therefore \cos\left(\dfrac{\pi}{3} + \dfrac{\pi}{6}\right) \neq \cos\dfrac{\pi}{3} + \cos\dfrac{\pi}{6}$

The addition and subtraction indentities for the primary trig functions are summarized below:

Addition Identities

$$\sin(x + y) = \sin x \cos y + \cos x \sin y$$
$$\cos(x + y) = \cos x \cos y - \sin x \sin y$$
$$\tan(x + y) = \frac{\tan x + \tan y}{1 - \tan x \tan y}$$

Subtraction Identities

$$\sin(x - y) = \sin x \cos y - \cos x \sin y$$
$$\cos(x - y) = \cos x \cos y + \sin x \sin y$$
$$\tan(x - y) = \frac{\tan x - \tan y}{1 + \tan x \tan y}$$

By substituting x = y in the addition identities, we get the so-called double angle identities:

Double Angle Identities

$$\sin 2x = 2\sin x \cos x$$
$$\cos 2x = \cos^2 x - \sin^2 x$$
$$\tan 2x = \frac{2\tan x}{1 - \tan^2 x}$$

Of course, the addition/subtraction/double angle identities are used to prove other identities.

EXAMPLE 6

Prove $\cos(x + y)\cos y + \sin(x + y)\sin y = \cos x$

Here, the L.S. is obviously much more complex than the R.S. Looking at this expression, the logical approach would be to substitute using addition formulas.

$$
\begin{aligned}
\text{L.S.} &= \cos(x + y)\cos y + \sin(x + y)\sin y \\
&= (\cos x\cos y - \sin x\sin y)\cos y + (\sin x\cos y + \cos x\sin y)\sin y \\
&= \cos x\cos^2 y - \sin x\sin y\cos y + \sin x\cos y\sin y + \cos x\sin^2 y \\
&= \cos x(\cos^2 y + \sin^2 y) \\
&= \cos x(1) \\
&= \cos x
\end{aligned}
$$

$\therefore \cos (x + y)\cos y + \sin (x + y)\sin y = \cos x$

Note the use of factoring in the fourth line of the proof, followed by substitution with a Pythagorean identity in the fifth line.

EXAMPLE 7

Prove $\tan x = \dfrac{\sin 2x}{1 + \cos 2x}$

Clearly, it is best to start with the R.S. here. The need to substitute with double angle identities should also be evident.

$$
\begin{aligned}
\text{R.S.} &= \frac{\sin 2x}{1 + \cos 2x} \\[2mm]
&\quad \frac{2\sin x\cos x}{1 + \cos^2 x - \sin^2 x} \\[2mm]
&= \frac{2\sin x\cos x}{2\cos^2 x} \quad \text{(using a Pythagorean identitiy and collecting like terms)} \\[2mm]
&= \frac{\sin x}{\cos x} \\[2mm]
&= \tan x
\end{aligned}
$$

$\therefore \tan x = \dfrac{\sin 2x}{1 + \cos 2x}$

PRACTICE EXERCISES

Prove each of the following identities:

1. $\cos\theta\csc\theta = \cot\theta$

2. $\sin\theta\tan\theta = \dfrac{\cos\theta}{\cot^2\theta}$

3. $(\sin^2\theta - \cos^2\theta)(\sin^2\theta + \cos^2\theta) = 2\sin^2\theta - 1$

4. $\sec^2 x\left(1 - \dfrac{\sin^2 x}{\cos^2 x}\right) = 2\sec^2 x - \sec^4 x$

5. $\sin(180° - x) + \cos(180° + x) + \cos(360° - x) = \tan(-x) + \tan(180° + x) - \sin(180° + x)$

6. $\tan x = \dfrac{\sin x + \sin^2 x}{\cos x + \cos x \sin x}$

7. $\sin(x + y)\sin(x - y) = \cos^2 y - \cos^2 x$

8. $(\cos x\cos y - \sin x\sin y)(\cos x\cos y + \sin x\sin y)$
   $= \cos^2 x + \cos^2 y - 1$

9. $\dfrac{\tan(x - y) + \tan y}{1 - \tan(x - y)\tan y} = \tan x$

10. $\dfrac{\cos 2x}{1 + \sin 2x} = \tan\left(\dfrac{\pi}{4} - x\right)$

Sample examination questions

1.

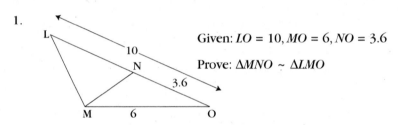

Given: $LO = 10, MO = 6, NO = 3.6$

Prove: $\triangle MNO \sim \triangle LMO$

2. Angle θ is in the second quadrant and $\csc\theta = \frac{11}{8}$. Find the primary trigonometric ratios.

3. Without using a calculator, evaluate each of the following, writing all steps:
a) $\cos 150°$ b) $\tan 1665°$ c) $\sin(-\frac{11\pi}{6})$

4. Evaluate without using a calculator:
$\sin^2 \frac{\pi}{3} \cos \frac{\pi}{3} - \tan \frac{\pi}{4} \sin \frac{\pi}{6}$

5. If $\sec\theta = -3.2194$ where $0° \leq \theta \leq 360°$, determine the value of θ, rounded to 1 decimal place, in the third quadrant.

6. Solve the following triangles:
a) $\triangle ABC$, $\angle A = 90°$, $a = 6.2$, $b = 3.3$
b) $\triangle KLM$, $\angle M = 36°$, $k = 87.9$, $l = 138.1$

7. A totem pole is standing at the top of a hill which has an angle of inclination of 9°. From the bottom of the hill, the

angle of elevation to the top of the pole is 21°. If the vertical height of the hill is 15 m, how tall is the totem pole?

8. Sketch the graph: $y = 2\cos(3\theta - \frac{\pi}{4}) - 1$, $-\pi \leq \theta \leq \pi$

9. Bonnie is riding on a swing which moves back and forth steadily like a pendulum.

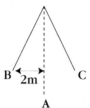

Bobbie measures her maximum horizontal distance at points B and C, from the center at A, as 2 meters. Starting his stopwatch at the moment Bonnie is passing through A, and counting each cycle through B, C and back to A as one, Bobbie observes her to complete five cycles in 10 seconds.

Write an equation for the horizontal distance as a function of time, in the form $y = a\sin k(t + c) + d$.

10. For the equation $(2\cos x - \sqrt{3})(\sin 2x + 1) = 0$:
    a) solve if $-\pi \leq x \leq \pi$.
    b) state the general solution.

11. Prove the following identities:
    a) $\cot\theta + \tan\theta = \csc\theta\sec\theta$
    b) $\dfrac{1 + \sin 2x}{\sin x + \cos x} = \sin x + \cos x$

(See page 107 for solutions)

Answers to practice exercises

PRACTICE EXERCISES, CHAPTER 1

1a) $\angle P = \angle S$ (given)
     $\angle Q = \angle SRT$ (given)
     $\triangle PQR \sim \triangle SRT$ (AA~)

b) $\angle D = \angle D$ (common)
     $EF \parallel GH$ (given)
     $\therefore \angle DEF = \angle DGH$ (PLT)
     $\therefore \triangle DEF \sim \triangle DGH$ (AA~)

c) $\dfrac{DO}{CA} = \dfrac{18}{3} = 6$

       $\dfrac{OG}{AT} = \dfrac{9}{1.5} = 6$

       $\dfrac{DG}{CT} = \dfrac{21}{3.5} = 6$

   $\therefore \dfrac{DO}{CA} = \dfrac{OG}{AT} = \dfrac{DG}{CT}$ (equality)

   $\therefore \triangle DOG \sim \triangle CAT$ (SSS~)

2a) $XY \parallel UV$ (given)
     $\therefore \angle XYZ = \angle VUZ$ (PLT)
     $\angle XZY = \angle VZU$ (OAT)
     $\therefore \triangle XYZ \sim \triangle VUZ$ (AA~)

b) $\dfrac{XZ}{2.25} = \dfrac{8}{3}$

     $\therefore XZ = 6$

3a) $\angle AGB = 50°$ (given)
     $\angle ADE = 50°$ (given)
     $\therefore \angle AGB = \angle ADE$ (equality)
     $\angle GAB = \angle DAE$ (common)
     $\therefore \triangle BAG \sim \triangle EAD$ (AA~)

b) $\dfrac{AB}{10} = \dfrac{24}{15}$

$\therefore AB = 16$

PRACTICE EXERCISES, CHAPTER 2

1. $\sin \theta = \dfrac{4}{5}, \cos \theta = \dfrac{3}{5}, \tan \theta = \dfrac{4}{3}$

2. let $x = $ the unknown side

$$x = \sqrt{(2)^2 + \left(\sqrt{21}\right)^2} = 5$$

$$\sin \beta = \dfrac{\sqrt{21}}{5}, \cos \beta = \dfrac{2}{5}, \tan \beta = \dfrac{\sqrt{21}}{2}$$

$$\csc \beta = \dfrac{5}{\sqrt{21}}, \sec \beta = \dfrac{5}{2}, \cot \beta = \dfrac{2}{\sqrt{21}}$$

3.

$x = \sqrt{(13)^2 - (5)2}$

$\quad = 12$

$\sin \theta = \dfrac{12}{13}, \cos \theta = \dfrac{5}{13}, \tan \theta = \dfrac{12}{5}$

$\csc \theta = \dfrac{13}{2}, \cot \theta = \dfrac{5}{12}$

4a) 1 　　　 b) $\dfrac{1}{2}$ 　　 c) $\sqrt{2}$ 　　 d) $\left(\dfrac{1}{2}\right)\left(\dfrac{\sqrt{3}}{2}\right)^2 = \dfrac{3}{8}$

e) $\left(\dfrac{1}{\sqrt{2}}\right)\left(\dfrac{1}{\sqrt{2}}\right) + \left(\dfrac{\sqrt{3}}{2}\right)\left(\sqrt{3}\right) = 2$

5a) 0.0523 　　 b) 0.1236 　 c) 0.6745 　 d) 0.2493

e) 8.8337 　　 f) 4.6942

6a) $\theta = 52.9°$ 　 b) $\alpha = 5.5°$ 　 c) $\lambda = 66.0°$ 　 d) $\mu = 16.0°$

e) $\eta = 30.0°$ 　 f) $\delta = 88.6°$

PRACTICE EXERCISES, CHAPTER 3

1

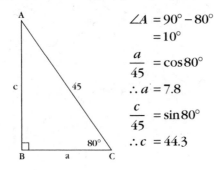

$$\angle A = 90° - 80°$$
$$= 10°$$

$$\frac{a}{45} = \cos 80°$$
$$\therefore a = 7.8$$

$$\frac{c}{45} = \sin 80°$$
$$\therefore c = 44.3$$

b)

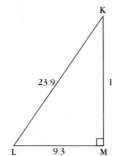

$$1 = \sqrt{(23.9)^2 - (9.3)^2}$$
$$= 22.0$$

$$\sin K = \frac{9.3}{23.9}$$
$$= 0.3891213$$
$$\therefore \angle K = 22.9°$$
$$\angle L = 90° - 22.9°$$
$$= 67.1°$$

2.

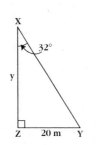

$$\tan 32° = \frac{20}{y}$$
$$\therefore y = \frac{20}{\tan 32°}$$
$$= 32.0$$

Therefore, y is 32.0 m long.

3.

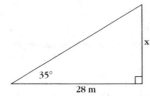

let x = the height of the flagpole

$$\frac{x}{28} = \tan 35°$$
$$\therefore x = 19.6$$

Therefore, the height of the flagpole is 19.6 m.

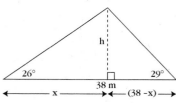

4.

let x = the height of the taller building

let y = the height of the shorter building

$$\frac{x}{15} = \tan 62.3°$$

$$\therefore x = 28.57079$$

let z = the difference in height between the two buildings

$$\frac{z}{15} = \tan 24.5°$$

$$\therefore z = 6.8358938$$

But $z = x - y$

$$\therefore y = x - z$$

$$= 28.57079 - 6.8358938$$

$$= 21.7348965.$$

Therefore the heights of the buildings are 28.6 m and 21.7 m.

5.

let h = the height of the tree

let x = the distance from the west point to the tree

$(38 - x)$ = the distance from the east point to the tree

$$\frac{h}{x} = \tan 26°$$

$$\therefore x = \frac{h}{\tan 26°}$$

$$\frac{h}{38 - x} = \tan 29°$$

$$h = 38\tan 29° - x\tan 29°$$

$$h - 38\tan 29° = -x\tan 29°$$

$$\therefore \frac{h}{\tan 26°} = \frac{38\tan 29° - h}{\tan 29°}$$

$$\therefore x = \frac{38\tan 29° - h}{\tan 29°}$$

$$h\tan 29° = \tan 26°(38\tan 29° - h)$$

$$h\tan 29° = 38\tan 26°\tan 29° - h\tan 26°$$

$$h(\tan 29° + \tan 26°) = 38\tan 26°\tan 29°$$

$$\therefore h = \frac{38\tan 26°\tan 29°}{\tan 29° + \tan 26°}$$

$$= \frac{10.273474}{1.0420416}$$

$$= 9.8589864$$

Therefore, the height of the tree is 9.9 m.

PRACTICE EXERCISES, CHAPTER 4

1. $r = \sqrt{(-4)^2 + (-7)^2} = \sqrt{65}$

      $\sin\theta = -\dfrac{7}{\sqrt{65}}, \cos\theta - \dfrac{4}{\sqrt{65}}, \tan\theta = \dfrac{7}{4}$

      $\csc\theta = -\dfrac{\sqrt{65}}{7}, \sec\theta = -\dfrac{\sqrt{65}}{4}, \cot\theta = \dfrac{4}{7}$

2.

$r = \sqrt{(-12)^2 + (-5)^2} = 13$

$\sin ß = -\dfrac{5}{3}, \cos ß - \dfrac{12}{13}$

$\csc ß = -\dfrac{13}{5}, \sec ß = -\dfrac{13}{12}, \cot ß = \dfrac{12}{5}$

3a) $\sin(180° - 45°) = \sin 45°$

         $= \dfrac{1}{\sqrt{2}}$

b) $\cos(180° - 45°) = -\cos 45°$

         $= -\dfrac{1}{\sqrt{2}}$

c) $\tan(180° - 30°) = -\tan 30°$

         $= -\dfrac{1}{\sqrt{3}}$

d) $\cos(180° + 60°) = -\cos 60°$

         $= -\dfrac{1}{2}$

e) $\cot(180° + 30°) = \cot 30°$

         $= \sqrt{3}$

f) $\sec(360° - 45°) = \sec 45°$

         $= \sqrt{2}$

g) $\sin(360° + 30°) = \sin 30°$

         $= \dfrac{1}{2}$

h) $780° \div 360° = 2 + 60°$

         remainder $\tan 60° = \sqrt{3}$

i) $\csc(360° + 60°) = \csc 60°$

         $= \dfrac{2}{\sqrt{3}}$

j) $\cos 120° = \cos(180° - 60°)$

         $= -\cos 60°$

         $= -\dfrac{1}{2}$

k) $1305° ÷ 360° = 3 + 225°R$

     $\cot 225° = \cot(180° + 45°)$

            $= \cot 45°$

            $= 1$

l) $1050° ÷ 360° = 2 + 330°R$

     $\sin 330° = \sin(360° -$

        $30°) = -\sin 30°$

          $= -\dfrac{1}{2}$

m) $-\sin 45° = -\dfrac{1}{\sqrt{2}}$

n) $\cos 150° = \cos(180° - 30°)$

            $= -\cos 30°$

            $= -\dfrac{\sqrt{3}}{2}$

o) $-\tan 240° = -\tan(180° + 60°)$

          $= -\tan 60°$

          $= -\sqrt{3}$

p) $\sec 300° = \sec(360° - 60°)$

          $= \sec 60°$

          $= 2$

q) $-\cot 135° = -\cot(180° - 45°)$

          $= \cot 45°$

          $= 1$

r) $930° ÷ 360° = 2 + 210°R$

     $-\csc 210° = -\csc(180° +$

        $30°) = \csc 30°$

          $= 2$

4a)   – 0.3746   b)   0.8391   c)   – 1.4267   d)   1.5399

  e)   – 0.7683   f)   5.1034

5a)   – 6.0°   b)   – 74.9°   c)   104.4°

6a)   b)

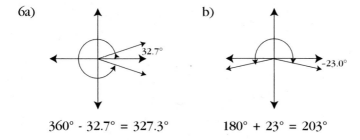

   360° - 32.7° = 327.3°   180° + 23° = 203°

c)

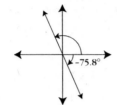

$$180° - 75.8° = 104.2°$$

7a)

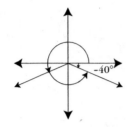

b)

180° − 114.4°   114.4°

$$\theta = 180° + 40°, 360° - 40°$$
$$= 220°, 320°$$

$$\theta = 114.4°, 180° + 65.6°$$
$$= 114.4°, 245.6°$$

PRACTICE EXERCISES, CHAPTER 5

1a)

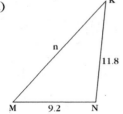

$$n^2 = (9.2)^2 + (11.8)^2 - 2(9.2)(11.8)\cos 125°$$
$$= 348.41492$$
$$\therefore n = 18.7$$

b)

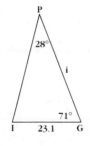

$$\angle I = 180° - (28° + 71°)$$
$$= 81°$$
$$\frac{i}{\sin 81°} = \frac{23.1}{\sin 28°}$$
$$\therefore i = \frac{23.1(\sin 81°)}{\sin 28°}$$
$$= 48.4$$

2a)

$$a^2 = m^2 + n^2 - 2mn\cos A$$

$$\cos A = \frac{a^2 - m^2 - n^2}{-2mn}$$

$$= \frac{(38.2)^2 - (49.1)^2 - (54)^2}{-2(49.1)(54)}$$

$$= \frac{-3867.57}{-5302.8}$$

$$= 0.7293448$$

$$\therefore \angle A = 43.2°$$

b)

$$\frac{8.1}{\sin T} = \frac{19.3}{\sin 105°}$$

$$\sin T = \frac{8.1(\sin 105°)}{19.3}$$

$$= 0.4053885$$

$$\therefore \angle T = 23.9°$$

3a)

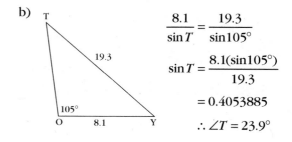

$$e^2 = d^2 + f^2 - 2df\cos E$$
$$= (10)^2 + (12)^2 - 2(10)(12)$$
$$\cos 36.3°$$
$$= 50.577212$$
$$\therefore e = 7.1$$

$$\sin D = \frac{10\sin 36.3°}{7.1}$$

$$= 0.8338213$$

$$\therefore \angle D = 56.5°$$

$$\angle F = 180° - (56.5° + 36.3°)$$

$$= 87.2°$$

b)

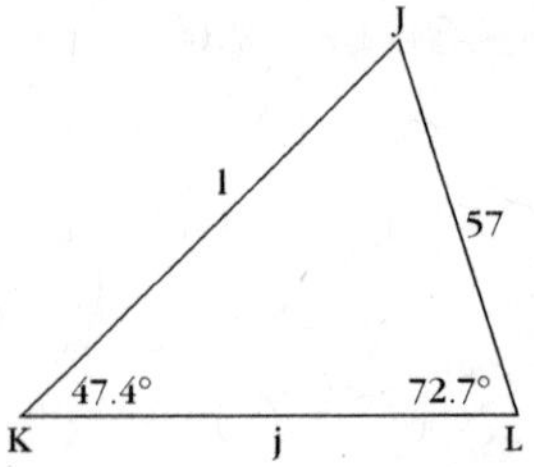

$$\angle J = 180° - (47.4° + 72.7°)$$
$$= 59.9°$$

$$\frac{j}{\sin 59.9°} = \frac{57}{\sin 47.4°}$$

$$\therefore j = \frac{57 \sin 59.9°}{\sin 47.4°}$$

$$= 67.0$$

$$\frac{1}{\sin 72.7°} = \frac{57}{\sin 47.4°}$$

$$\therefore 1 = \frac{57 \sin 72.7°}{\sin 47.4°}$$

$$= 73.9$$

c)

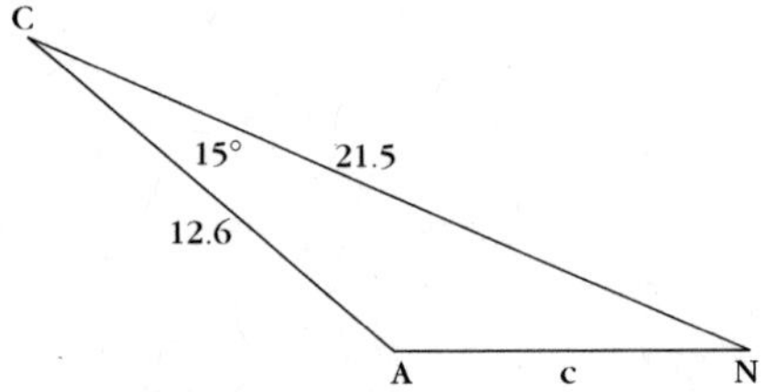

$$c^2 = a^2 + n^2 - 2an \cos C$$

$$= (21.5)^2 + (12.6)^2 - 2(21.5)(12.6) \cos 15°$$

$$= 97.671387$$

$$\therefore c = 9.9$$

$$\frac{12.6}{\sin N} = \frac{9.9}{\sin 15°}$$

$$\sin N = \frac{12.6 \sin 15°}{9.9}$$

$$= 0.329406$$

$$\therefore \angle N = 19.2°$$

$$\angle A = 180° - (15° + 19.2°)$$

$$= 145.8°$$

1.

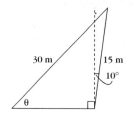

let θ = the angle the rope makes
 with the ground

$$\frac{15}{\sin\theta} = \frac{30}{\sin100°}$$

$$\sin\theta = \frac{15\sin100°}{30}$$

$$= 0.4924038$$

$$\therefore \theta = 29.5°$$

Therefore, the rope makes an angle of 29.5° with the ground.

2.

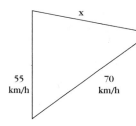

let x = the distance between the motorcyclists

After 2 h, the motorcyclist travelling 55 km/h will have gone 2(55) = 110 km, while the other will have travelled 2(70) = 140 km.

$$x^2 = (110)^2 + (140)^2 - 2(110)(140)\cos53.2°$$
$$= 13250.073$$
$$\therefore x = 115.1$$

Therefore, after 2 h, the motorcyclists will be 115 km apart.

3.

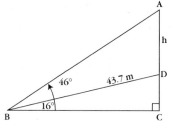

let b = the tower's height
$\angle A = 90° - 46°$
$\quad = 44°$
$\angle ABD = 46° - 16°$
$\quad\quad = 30°$

in ΔABD, $\dfrac{b}{\sin30°} = \dfrac{43.7}{\sin44°}$

$$\therefore b = \frac{43.7\sin30°}{\sin44°}$$

$$= 31.45431$$

Therefore, the height of the tower is 31.5 m.

4.

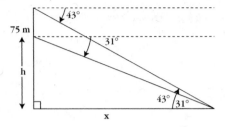

let x = the distance from the launch site to the pond

let b = the height of the balloon at Linda's first sighting of the pond

$$\frac{b}{x} = \tan 31° \qquad\qquad \frac{b+75}{x} = \tan 43°$$

$$b = x\tan 31° \qquad\qquad b = x\tan 43° - 75$$

$$\therefore x\tan 31° = x\tan 43° - 75$$

$$x(\tan 31° - \tan 43°) = -75$$

$$\therefore x = \frac{-75}{\tan 31° - \tan 43°}$$

$$= 226.13897$$

Therefore, the pond is about 226 m from the launch site.

5. In $\triangle BKN$: $\angle B = 180° - (27.8° + 58.2°)$
$$= 94°$$

$$\frac{k}{\sin 27.8°} = \frac{15.2}{\sin 94°} \qquad\qquad \frac{n}{\sin 58.2°} = \frac{15.2}{\sin 94°}$$

$$\therefore k = \frac{15.2\sin 27.8°}{\sin 94°} \qquad\qquad \therefore n = \frac{15.2\sin 58.2°}{\sin 94°}$$

$$= 7.1063877 \qquad\qquad = 12.949914$$

In $\triangle AKB$: $\quad \frac{b}{n} = \tan 23.2°$

$$\therefore b = 12.949914\tan 23.2°$$

$$= 5.55$$

In $\triangle ANB$: $\quad \frac{b}{k} = \tan 35.8°$

$$\therefore b = 5.1252885$$

Therefore, Jake determined the height of the airplane to be about 5.6 km, while Janet figured it to be about 5.1 km.

PRACTICE EXERCISES, CHAPTER 7

1a) $\dfrac{\pi}{36}$ b) $\dfrac{\pi}{5}$ c) $-\dfrac{35\pi}{18}$ d) $\dfrac{3\pi}{2}$ e) $\dfrac{\pi}{9}$

2a) $22.5°$ b) $108°$ c) $-540°$ d) $15°$ e) $-450°$

3.

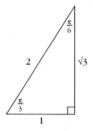

4a) $\dfrac{1}{\sqrt{2}}$ b) $\dfrac{1}{2}$ c) $\dfrac{1}{\sqrt{3}}$ d) $\dfrac{\sqrt{3}}{2}$

e) $\tan\left(\pi - \dfrac{\pi}{4}\right) = -\tan\dfrac{\pi}{4}$

$\qquad = -1$

f) $\cos\left(\pi - \dfrac{\pi}{3}\right) = -\cos\dfrac{\pi}{3}$

$\qquad = -\dfrac{1}{2}$

g) $\sin\left(\pi - \dfrac{\pi}{6}\right) = \sin\dfrac{\pi}{6}$

$\qquad = \dfrac{1}{2}$

h) $\cos\left(\pi + \dfrac{\pi}{3}\right) = -\cos\dfrac{\pi}{3}$

$\qquad = -\dfrac{1}{2}$

i) $\tan\left(2\pi - \dfrac{\pi}{6}\right) = -\tan\dfrac{\pi}{6}$

$\qquad = -\dfrac{1}{\sqrt{3}}$

j) $\cos\left(\pi + \dfrac{\pi}{6}\right) = -\cos\dfrac{\pi}{6}$

$\qquad = -\dfrac{\sqrt{3}}{2}$

k) $\sin\left(\pi + \dfrac{\pi}{4}\right) = -\sin\dfrac{\pi}{4}$

$\qquad = -\dfrac{1}{\sqrt{2}}$

l) $-\tan\left(2\pi - \dfrac{\pi}{4}\right) = \tan\dfrac{\pi}{4}$

$\qquad = 1$

5a) 0.9659 b) 3.0777 c) -0.6119 d) -0.4161 e) 1.1395

6a) 0.5106 b) 0.7854 c) 0.7854 d) -0.7649 e) 0.4498

1. $y = \cos\theta$

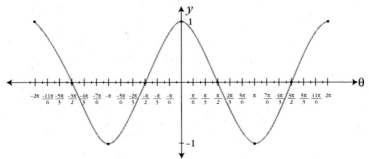

2.

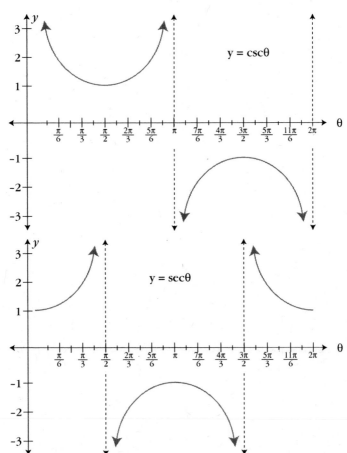

$y = \csc\theta$

$y = \sec\theta$

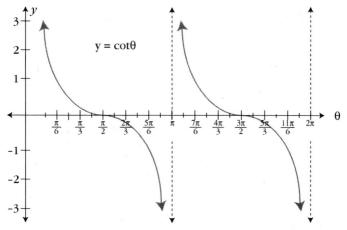

3.

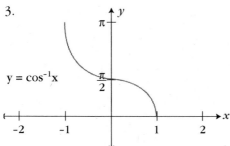

PRACTICE EXERCISES, CHAPTER 9

1a)　amplitude = 2

$$\text{period} = \frac{2\pi}{4} = \frac{\pi}{2}$$

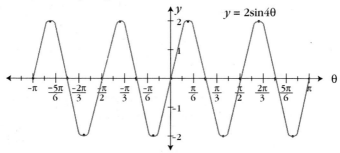

97

b) amplitude = 3

   period = π

   phaseshift = $\dfrac{\pi}{4}$ right

   vertical translation = 1 up

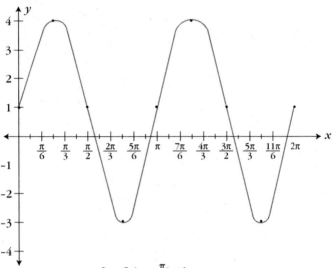

$$y = 3\cos 2\left(x - \tfrac{\pi}{4}\right) + 1$$

c) $y = \dfrac{1}{2}\sin 3\left(x + \dfrac{\pi}{6}\right) - 2$

   amplitude = $\dfrac{1}{2}$

   period = $\dfrac{2\pi}{3}$

   phaseshift = $\dfrac{\pi}{6}$ left

   vertical translation = 2 down

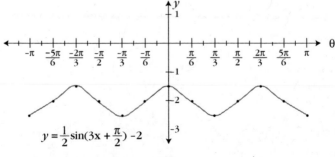

$$y = \dfrac{1}{2}\sin\left(3x + \tfrac{\pi}{2}\right) - 2$$

98

2a) maximum depth = 20 m, minimum depth = 8 m

middle depth = $\dfrac{20+8}{2}=14$ m

$\therefore$ amplitude $= 20-14 = 6$ m

period $= 12$ h

$\therefore \dfrac{2\pi}{k}=12$

$k=\dfrac{\pi}{6}$

let d = the depth of the water in meters

let t = the time elapsed in hours

at $t = 0$, $d = 20$ (maximum height)

So, we can express depth using a cosine function.

Therefore, d = $6\cos\dfrac{\pi}{6}t + 14$ is the required equation.

b)

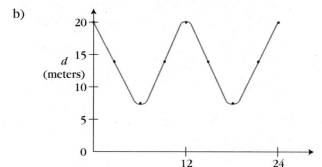

3a) at $t = 0, h = 12\sin 2\pi\left(0-\dfrac{1}{4}\right)+17$

$= 12\sin\left(-\dfrac{\pi}{2}\right)+17$

$= 5$

Therefore, the height of Susie's foot is 5 cm.

b) 12 cm

c) period $= \dfrac{2\pi}{k}=\dfrac{2\pi}{2\pi}=1$

Therefore, it takes 1 second for her foot to make one rotation.

d)

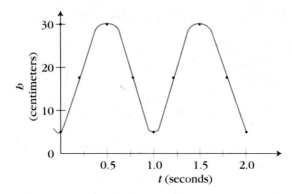

e) At t = 0, 1, 2, etc., her other foot will be the maximum height (29 cm) above the ground.

Thus, we can use a cosine function with no phase shift to represent the other foot's position.

$\therefore b = 12\cos 2\pi t + 17$ is the required equation.

PRACTICE EXERCISES, CHAPTER 10

1a)

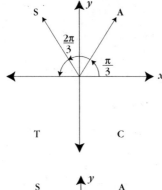

$a = \dfrac{\pi}{3}$

from 0 to π, $x = \dfrac{\pi}{3}, \dfrac{2\pi}{3}$

from 0 to $-\pi$, no solution

$\therefore x = \dfrac{\pi}{3}, \dfrac{2\pi}{3}$ is the solution.

b)

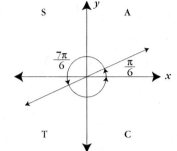

$$a = \frac{\pi}{6}$$

from 0 to 2π

$$x = \frac{\pi}{6}, \frac{7\pi}{6}$$

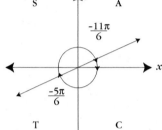

from 0 to -2π

$$x = -\pi + \frac{\pi}{6}, -2\pi + \frac{\pi}{6}$$

$$= -\frac{5\pi}{6}, -\frac{11\pi}{6}$$

$$\therefore x = -\frac{11\pi}{6}, \frac{5\pi}{6}, \frac{\pi}{6}, \frac{7\pi}{6}$$
is the solution

c)

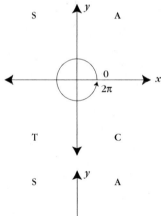

$$-2\pi \leq 2x \leq 2\pi$$

$$a = 0$$

from 0 to 2π:

$$2x = 0, 2\pi$$

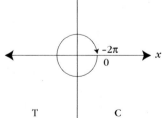

from 0 to -2π

$$2x = 0, -2\pi$$

$$2x = -2\pi, 0, 2\pi$$

$$\therefore x = -\pi, 0, \pi$$
is the solution

d)

$$\cos x = \frac{1}{2}$$ or $$\tan x = -1$$

$$a = \frac{\pi}{3}$$ $$a = \frac{\pi}{4}$$ but we want x for a negative trig ratio

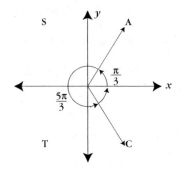

$$x = \frac{\pi}{3}, 2\pi - \frac{\pi}{3}$$ $$x = \frac{3\pi}{4}, 2\pi - \frac{\pi}{4}$$

$$= \frac{\pi}{3}, \frac{5\pi}{3}$$ $$= \frac{3\pi}{4}, \frac{7\pi}{4}$$

$$\therefore x = \frac{\pi}{3}, \frac{3\pi}{4}, \frac{5\pi}{3}, \frac{7\pi}{4} \text{ is the solution.}$$

e) $\cos 2x(\cos 2x + 1) = 0$, $-4\pi \leq 2x \leq 4\pi$

$$\cos 2x = 0$$ or $$\cos 2x = -1$$

$$a = \frac{\pi}{2}$$ $$a = 0 \text{ but we want } 2x \text{ for a negative trig ratio}$$

from 0 to 4π: from 0 to 4π:

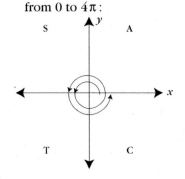

$$2x = \frac{\pi}{2}, \frac{3\pi}{2}, \frac{5\pi}{2}, \frac{7\pi}{2}$$ $$2x = \pi, 3\pi$$

102

from 0 to –4π:

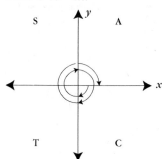

from 0 to –4π:

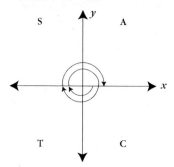

$$2x = -\frac{\pi}{2}, -\frac{3\pi}{2}, -\frac{5\pi}{2}, -\frac{7\pi}{2} \qquad 2x = -\pi, -3\pi$$

$$2x = -\frac{7\pi}{2}, -3\pi, -\frac{5\pi}{2}, -\frac{3\pi}{2}, -\pi, -\frac{\pi}{2}, \frac{\pi}{2}, \pi, \frac{3\pi}{2}, \frac{5\pi}{2}, 3\pi, \frac{7\pi}{2}$$

$$\therefore x = --\frac{7\pi}{4}, -\frac{3\pi}{2}, -\frac{5\pi}{4}, -\frac{3\pi}{4}, -\frac{\pi}{2}, -\frac{\pi}{4}, \frac{\pi}{4}, \frac{\pi}{2}, \frac{3\pi}{4}, \frac{5\pi}{4}, \frac{3\pi}{2}, \frac{7\pi}{4}$$

2a)

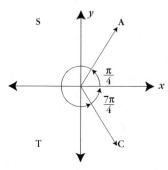

$a = \dfrac{\pi}{4}$

The general solution is:

$x = \dfrac{\pi}{4} + 2\pi n$ or $x = \dfrac{7\pi}{4} + 2\pi n$, $n \in I$

b)

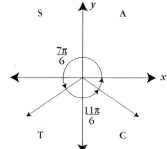

$a = \dfrac{\pi}{6}$ but we want $2x$ for a negative trig ratio

$2x = \dfrac{7\pi}{6} + 2\pi n$ or $2x = \dfrac{11\pi}{6} + 2\pi n$

Therefore, the general solution is:

$x = \dfrac{7\pi}{12} + \pi n$ or $x = \dfrac{11\pi}{12} + \pi n$

$n \in I$

c)

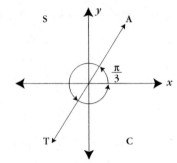

$$a = \frac{\pi}{3}$$

$$2x = \frac{\pi}{3} + \pi n$$

Therefore, the general solution is:

$$x = \frac{\pi}{6} + \frac{\pi}{2} n, \ n \in I$$

PRACTICE EXERCISES, CHAPTER 11

1. $L.S. = \cos\theta \csc\theta$

    $= \cos\theta . \dfrac{1}{\sin\theta}$

    $= \cot\theta$

    $\therefore \cos\theta \csc\theta = \cot\theta$

2. $R.S. = \dfrac{\cos\theta}{\cot^2\theta}$

    $= \dfrac{\cos\theta}{\dfrac{\cos^2\theta}{\sin^2\theta}}$

    $= \cos\theta . \dfrac{\sin^2}{\cos^2\theta}$

    $= \sin\theta . \dfrac{\sin\theta}{\cos\theta}$

    $= \sin\theta \tan\theta$

    $\therefore \sin\theta \tan\theta = \dfrac{\cos\theta}{\cot^2\theta}$

3. $L.S. = (\sin^2\theta - \cos^2\theta)(\sin^2\theta + \cos^2\theta)$

    $= (\sin^2\theta - \cos^2\theta)(1)$

    $= \sin^2\theta - (1 - \sin^2\theta)$

    $= 2\sin^2\theta - 1$

    $\therefore (\sin^2\theta - \cos^2\theta)(\sin^2\theta + \cos^2\theta) = 2\sin^2\theta - 1$

4. $L.S. = \sec^2 x\left(1 - \dfrac{\sin^2 x}{\cos^2 x}\right)$ $\qquad$ $R.S. = 2\sec^2 x - \sec^4 x$

$\qquad = \dfrac{1}{\cos^2 x}\left(\dfrac{\cos^2 x - \sin^2 x}{\cos^2 x}\right)$ $\qquad = \dfrac{2}{\cos^2 x} - \dfrac{1}{\cos^4 x}$

$\qquad = \dfrac{\cos^2 x - (1 - \cos^2 x)}{\cos^4 x}$ $\qquad = \dfrac{2\cos^2 x - 1}{\cos^4 x}$

$\qquad = \dfrac{2\cos^2 x - 1}{\cos^4 x}$

$\therefore \sec^2 x\left(1 - \dfrac{\sin^2 x}{\cos^2 x}\right) = 2\sec^2 x - \sec^4 x$

5. $L.S. = \sin(180° - x) + \cos(180° + x) + \cos(360° - x)$
$\qquad = \sin x + (-\cos x) + \cos x$
$\qquad = \sin x$
$R.S. = \tan(-x) + \tan(180° + x) - \sin(180° + x)$
$\qquad = -\tan x + \tan x - (-\sin x)$
$\qquad = \sin x$
$\therefore \sin(180° - x) + \cos(180° + x) + \cos(360° - x)$
$= \tan(-x) + \tan(180° + x)$

6. $R.S. = \dfrac{\sin x + \sin^2 x}{\cos x + \cos x \sin x}$

$\qquad = \dfrac{\sin x(1 + \sin x)}{\cos x(1 + \sin x)}$

$\qquad = \tan x$

$\therefore \tan x = \dfrac{\sin x + \sin^2 x}{\cos x + \cos x \sin x}$

7. $L.S. = \sin(x + y)\sin(x - y)$
$\qquad = (\sin x\cos y + \cos x\sin y)(\sin x\cos y - \cos x\sin y)$
$\qquad = \sin^2 x\cos^2 y - \sin x\cos x\sin y\cos y + \sin x\cos x\sin y\cos y - \cos^2 x\sin^2 y$
$\qquad = \sin^2 x\cos^2 y - \cos^2 x\sin^2 y$
$\qquad = \cos^2 y(1 - \cos^2 x) - \cos^2 x(1 - \cos^2 y)$
$\qquad = \cos^2 y - \cos^2 x\cos^2 y - \cos^2 x + \cos^2 x\cos^2 y$

105

$$= \cos^2 y - \cos^2 x$$

$$\therefore \sin(x + y) \sin(x - y) = \cos^2 y - \cos^2 x$$

8.   L.S. $= (\cos x \cos y - \sin x \sin y)(\cos x \cos y + \sin x \sin y)$

$$= \cos^2 x \cos^2 y - \sin^2 x \sin^2 y$$

$$= (1 - \sin^2 x)\cos^2 y - (1 - \cos^2 x)\sin^2 y$$

$$= \cos^2 y - \sin^2 x \cos^2 y - \sin^2 y + \cos^2 x \sin^2 y$$

$$= \cos^2 y - (1 - \cos^2 x)(1 - \sin^2 y) - \sin^2 y + \cos^2 x \sin^2 y$$

$$= \cos^2 y - (1 - \sin^2 y - \cos^2 x + \cos^2 x \sin^2 y) - \sin^2 y + \cos^2 x \sin^2 y$$

$$= \cos^2 y - 1 + \sin^2 y + \cos^2 x - \cos^2 x \sin^2 y - \sin^2 y + \cos^2 x \sin^2 y$$

$$= \cos^2 x + \cos^2 y - 1$$

$$\therefore (\cos x \cos y - \sin x \sin y)(\cos x \cos y + \sin x \sin y)$$

$$= \cos^2 x + \cos^2 y - 1$$

9.   L.S. $= \dfrac{\tan(x - y) + \tan y}{1 - \tan(x - y)\tan y}$

$$= \dfrac{\dfrac{\tan x - \tan y}{1 + \tan x \tan y} + \tan y}{1 - \dfrac{\tan x - \tan y}{1 + \tan x \tan y}.\tan y}$$

$$= \dfrac{\dfrac{\tan x - \tan y + \tan y + \tan x \tan^2 y}{1 + \tan x \tan y}}{\dfrac{1 + \tan x \tan y - (\tan x \tan y - \tan^2 y)}{1 + \tan x \tan y}}$$

$$= \dfrac{\tan x + \tan x \tan^2 y}{1 + \tan^2 y}$$

$$= \dfrac{\tan x(1 + \tan^2 y)}{1 + \tan^2 y}$$

$$= \tan x$$

$$\therefore \dfrac{\tan(x - y) + \tan y}{1 - \tan(x - y)\tan y} = \tan x$$

10. L.S. $= \dfrac{\cos 2x}{1+\sin 2x}$

   R.S. $= \tan\left(\dfrac{\pi}{4} - x\right)$

$= \dfrac{\cos^2 x - \sin^2 x}{1 + 2\sin x \cos x}$

$= \dfrac{\tan\dfrac{\pi}{4} - \tan x}{1 + \tan\dfrac{\pi}{4}\tan x}$

$= \dfrac{\cos^2 x - \sin^2 x}{\sin^2 x + \cos^2 x + 2\sin x \cos x}$

$= \dfrac{1 - \tan x}{1 + \tan x}$

$= \dfrac{(\cos x - \sin x)(\cos x + \sin x)}{(\cos x + \sin x)(\cos x + \sin x)}$

$= 1 - \dfrac{\dfrac{\sin x}{\cos x}}{1 + \dfrac{\sin x}{\cos x}}$

$= \dfrac{\cos x - \sin x}{\cos x + \sin x}$

$= \dfrac{\dfrac{\cos x - \sin x}{\cos x}}{\dfrac{\cos x + \sin x}{\cos x}}$

$= \dfrac{\cos x - \sin x}{\cos x + \sin x}$

$\therefore \dfrac{\cos 2x}{1 + \sin 2x} = \tan\left(\dfrac{\pi}{4} - x\right)$

Solutions to sample examination questions

1.
$$\frac{MO}{LO} = \frac{6}{10} = 0.6$$

$$\frac{NO}{MO} = \frac{3.6}{6} = 0.6$$

$$\therefore \frac{MO}{LO} = \frac{NO}{MO} \text{ (by equality)}$$

$\angle MON = \angle LOM$ (common)

$\therefore \Delta MON \sim \Delta LMO(SAS \sim)$

2.

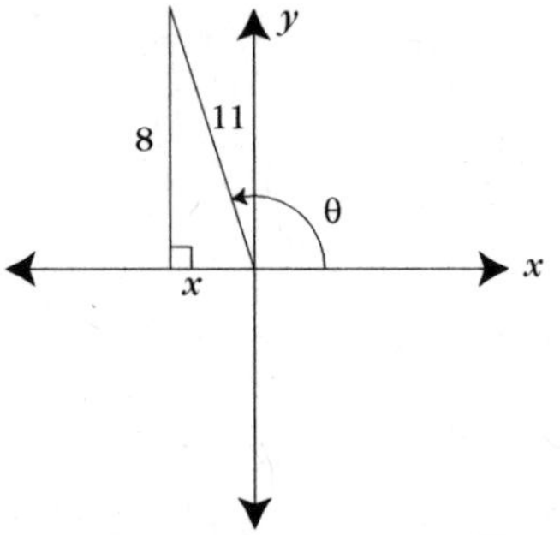

$$x^2 = (11)^2 - (8)^2$$
$$= 57$$
$$\therefore x = -\sqrt{57}$$

$$\therefore \sin\theta = \frac{8}{11}, \cos\theta = -\frac{\sqrt{57}}{11}, \tan\theta = -\frac{8}{\sqrt{57}}$$

3. a) $\cos 150° = \cos(180° - 30°)$
$$= -\cos 30°$$

$$= -\frac{\sqrt{3}}{2}$$

b) $1665° \div 360° = 4 + 225° R$
$\tan 225° = \tan(180° + 45°)$
$$= \tan 45°$$
$$= 1$$

3. c) $\sin\left(-\frac{11\pi}{6}\right) = -\sin\frac{11\pi}{6}$

$$= -\sin\left(2\pi - \frac{\pi}{6}\right)$$

$$= \sin\frac{\pi}{6}$$

$$= \frac{1}{2}$$

4. $\sin^2\frac{\pi}{3}\cos\frac{\pi}{3} - \tan\frac{\pi}{4}\sin\frac{\pi}{6} = \left(\frac{\sqrt{3}}{2}\right)^2\left(\frac{1}{2}\right) - (1)\left(\frac{1}{2}\right)$

$$= \frac{3}{8} - \frac{1}{2}$$

$$= -\frac{1}{8}$$

5.

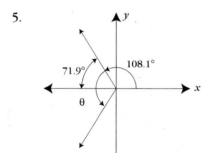

$\cos\theta = -0.3106168$

$\theta = 108.1°$ according to the calculator but, we want θ in the third quadrant.

$\therefore \theta = 180° + 71.9°$

$\quad = 251.9°$

6 a)

$\cos C = \frac{3.3}{6.2}$

$\quad = 0.532258$

$\therefore \angle C = 57.8°$

$\angle B = 90° - 57.8° = 32.2°$

$c = \sqrt{(6.2)^2 - (3.3)^2}$

$\quad = 5.2$

b) $m^2 = (87.9)^2 + (138.1)^2 - 2(87.9)(138.1)\cos 36°$

$\qquad = 7156.7216$

$\qquad \therefore m = 84.6$

$\dfrac{87.9}{\sin K} = \dfrac{84.6}{\sin 36°}$

$\sin K = \dfrac{87.9 \sin 36°}{84.6}$

$\qquad = 0.610713$

$\qquad \therefore \angle K = 37.6°$

$\angle L = 180° - (37.6° + 36°)$

$\qquad = 106.4°$

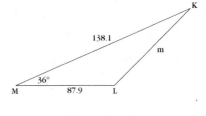

7.

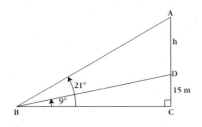

let b = the height of the totem pole

In $\triangle DBC$, $\dfrac{15}{BD} = \sin 9°$

$\qquad \therefore BD = \dfrac{15}{\sin 9°}$

$\qquad\qquad = 95.886798$

$\angle ABD = 21° - 9° = 12°$

$\angle ADB = \angle DBC + \angle DCB$

$\qquad\qquad = 9° + 90°$

$\qquad\qquad = 99°$

$\angle BAD = 180° - (99° + 12°)$

$\qquad\qquad = 69°$

$\dfrac{b}{\sin 12°} = \dfrac{95.886798}{\sin 69°}$

$\qquad \therefore b = 21.4$

Therefore, the totem pole is 21.4 m tall.

8. amplitude = 2

period $= \dfrac{2\pi}{3}$

phase shift $= \dfrac{\pi}{12}\,right$

vertical translation = 1 down

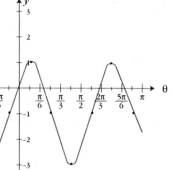

9. amplitude = 2 m

period $= \dfrac{5}{10} = 0.5$ second (1 cycle takes 0.5 seconds)

$$\dfrac{2\pi}{k} = 0.5$$

$$\therefore k = \dfrac{2\pi}{0.5} = 4\pi$$

Therefore, the required equation is:

$$h = 2\sin 4\pi t$$

10 a)

$\cos x = \dfrac{\sqrt{3}}{2}$ or $\sin 2x = -1$

$a = \dfrac{\pi}{6}$ $-2\pi \le 2x \le 2\pi$

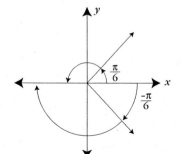 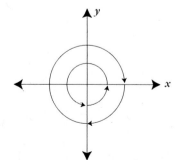

$$x = -\frac{\pi}{6}, \frac{\pi}{6} \qquad\qquad 2x = \frac{3\pi}{2}, -\frac{\pi}{2}$$

$$x = \frac{3\pi}{4}, -\frac{\pi}{4}$$

$\therefore x = -\dfrac{\pi}{4}, -\dfrac{\pi}{6}, \dfrac{\pi}{6}, \dfrac{3\pi}{4}$ is the solution

b)
$$2x = \frac{3\pi}{2} + 2\pi n$$

$$x = \frac{3\pi}{4} + \pi n$$

Therefore, the general solution is:

$$x = -\frac{\pi}{6} + 2\pi n, x = \frac{\pi}{6} + 2\pi n, x = \frac{3\pi}{4} + \pi n \text{ where } n \in I$$

11a) $L.S. = \cot\theta + \tan\theta$ $\qquad\qquad$ $R.S. = \csc\theta\sec\theta$

$\qquad\quad = \dfrac{\cos\theta}{\sin\theta} + \dfrac{\sin\theta}{\cos\theta}$ $\qquad\qquad\qquad = \dfrac{1}{\sin\theta\cos\theta}$

$\qquad\quad = \dfrac{\cos^2\theta + \sin^2\theta}{\sin\theta\cos\theta}$

$\qquad\quad = \dfrac{1}{\sin\theta\cos\theta}$

$\therefore \cot\theta + \tan\theta = \csc\theta\sec\theta$

b) $L.S. = \dfrac{1 + \sin 2x}{\sin x + \cos x}$

$\qquad\quad = \dfrac{1 + 2\sin x\cos x}{\sin x + \cos x}$

$\qquad\quad = \dfrac{\sin^2 x + \cos^2 x + 2\sin x\cos x}{\sin x + \cos x}$

$\qquad\quad = \dfrac{(\sin x + \cos x)(\sin x + \cos x)}{\sin x + \cos x}$

$\qquad\quad = \sin x + \cos x$

$\therefore \dfrac{1 + \sin 2x}{\sin x + \cos x} = \sin x + \cos x$

Memory joggers

AAA~	$\sin\theta = \dfrac{Opp}{Hyp} = \dfrac{y}{r}$	$\csc\theta = \dfrac{1}{\sin\theta}$
AA~		
SAS~	$\cos\theta = \dfrac{Adj}{Hyp} = \dfrac{x}{r}$	$\sec\theta = \dfrac{1}{\cos\theta}$
SSS~	$\tan\theta = \dfrac{Opp}{Hyp} = \dfrac{y}{x}$	$\cot\theta = \dfrac{1}{\tan\theta}$

SOHCAHTOA

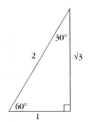

CAST

$\sin(180° - a) = \sin a$ $\qquad$ $\sin(180° + a) = -\sin a$

$\cos(180° - a) = -\cos a$ $\qquad$ $\cos(180° + a) = -\cos a$

$\tan(180° - a) = -\tan a$ $\qquad$ $\tan(180° + a) = \tan a$

$\sin(360° - a) = -\sin a$ $\qquad$ $\sin(-a) = -\sin a$

$\cos(360° - a) = \cos a$ $\qquad$ $\cos(-a) = \cos a$

$\tan(360° - a) = -\tan a$ $\qquad$ $\tan(-a) = -\tan a$

$c^2 = a^2 + b^2 - 2ab\cos C$

$\dfrac{a}{\sin A} = \dfrac{b}{\sin B} = \dfrac{c}{\sin C}$

$180° = \pi$ radians

$y = a\sin k(\theta + c) + d$

$y = a\cos k(\theta + c) + d$

$\tan\theta = \dfrac{\sin\theta}{\cos\theta}$ $\qquad\qquad$ $\cot\theta = \dfrac{\cos\theta}{\sin\theta}$

$\sin^2\theta + \cos^2\theta = 1$
$1 + \tan^2\theta = \sec^2\theta$
$1 + \cot^2\theta = \csc^2\theta$

$\sin(x + y) = \sin x \cos y + \cos x \sin y$ $\sin(x - y) = \sin x \cos y - \cos x \sin y$
$\cos(x + y) = \cos x \cos y - \sin x \sin y$ $\cos(x - y) = \cos x \cos y + \sin x \sin y$

$$\tan(x + y) = \frac{\tan x + \tan y}{1 - \tan x \tan y} \qquad \tan(x - y) = \frac{\tan x - \tan y}{1 + \tan x \tan y}$$

$\sin 2x = 2\sin x \cos x$
$\cos 2x = \cos^2 x - \sin^2 x$
$$\tan 2x = \frac{2\tan x}{1 - \tan^2 x}$$

NOTES & UPDATES

NOTES & UPDATES

NOTES & UPDATES

NOTES & UPDATES

NOTES & UPDATES